THE OXFORD BOOK OF INSECTS

THE OXFORD BOOK
OF INSECTS

Illustrations by

JOYCE BEE, DEREK WHITELEY,

[PETER PARKS

Text by

[J]HN BURTON

with

[...]ROW   A. A. ALLEN

[...]ER   I. LANSBURY

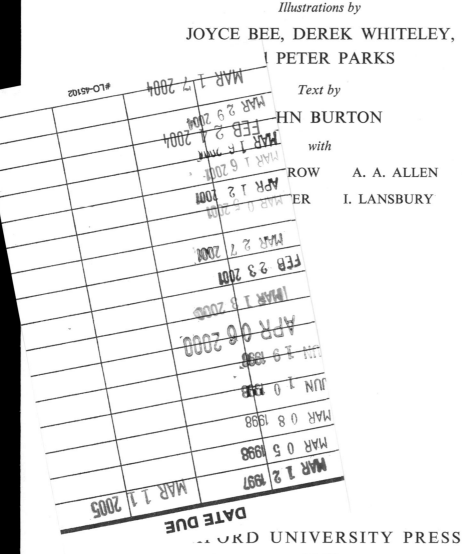
[...]ORD UNIVERSITY PRESS

1968

Oxford University Press, Ely House, London W.1

GLASGOW NEW YORK TORONTO MELBOURNE WELLINGTON
CAPE TOWN SALISBURY IBADAN NAIROBI LUSAKA ADDIS ABABA
BOMBAY CALCUTTA MADRAS KARACHI LAHORE DACCA
KUALA LUMPUR HONG KONG TOKYO

Printed in Great Britain by Jesse Broad & Co. Ltd., Old Trafford, Manchester

Contents

ACKNOWLEDGEMENTS

The 780 insects illustrated in this book have been drawn, where possible, from live insects, and otherwise from a combination of preserved specimens, mainly from the Hope Dept. of Entomology, University Museum, Oxford, and colour photographs. We are grateful, especially, to Mr. E. Taylor of the Hope Dept. for finding and checking material in Oxford; to the Dept. of Entomology, British Museum (Natural History) for making available mounted specimens; to Mr. Robert Goodden of Worldwide Butterflies Ltd. for providing both living and mounted specimens of Lepidoptera; and to Dr. D. R. Ragge for enabling us to refer to photographs in the national collection at the British Museum (Natural History) taken by Dr. C. G. Butler, Mr. R. A. Farrow, Mr. and Mrs. R. G. Foord, and Mr. J. A. Grant.

We wish to thank Dr. B. M. Hobby for his help and encouragement over the whole book, but especially on the Lepidoptera section, and Dr. T. R. E. Southwood for help over a problem with the Hemiptera. We are grateful to the librarian of the Royal Entomological Society for help in obtaining the specialist books on which any entomologist must greatly depend in preparing a general book of this kind, and also to the late Mr. R. B. Benson, Mr. J. D. Bradley, Mr. D. S. Fletcher, Mr. E. W. Groves, Mr. D. E. Kimmins, Mr. P. N. Lawrence, Mr. C. Moreby, Mr. J. Quinlan, Dr. D. R. Ragge, and Mr. P. H. Ward for helpful and up-to-date information on the insects on which they are specialists.

We have been very glad of the specialist help in writing the text pages of Mr. A. A. Allen (Coleoptera), Dr. I. H. H. Yarrow (Hymenoptera), Mr. L. Parmenter (Diptera), and Mr. Ivor Lansbury (Hemiptera). The plates were drawn by Miss Joyce Bee (Lepidoptera, pp. 41–121), Mr. Derek Whiteley (pp. 1–31 and pp. 143–191), and Mr. Peter Parks (pp. 33–39 and pp. 123–141). The jacket was designed and drawn by Mr. Derek Whiteley. They and we appreciate the skill and care of the plate-makers in reproducing their work so faithfully.

INTRODUCTION

This book belongs to the same series as the *Oxford Books of Wild Flowers, Garden Flowers, Flowerless Plants,* and *Birds.* It is an introduction to the insects of the British Isles, and its main purpose is to encourage people to observe and identify the vast range of very varied insects to be found around them.

Nearly 800 different species are illustrated in this book, and even more are included in the opposite pages of text; but this is still a small proportion of the 20,000 or more species which exist in Britain. Therefore this book has presented a problem of selection. Examples of all the twenty-five orders of British insects have been given, though some of the less-well-known orders are represented by only one or two species. The insects which attract the most attention from the layman, such as butterflies and moths, dragonflies and grasshoppers, are given a relatively generous amount of space, and, though many very small insects are shown in this book, often because they are important pests, microscopic species are not included. The greater majority of those species shown are common and widespread, though a few rarities are included because, like the Large Blue or Camberwell Beauty Butterflies, they are well known or especially interesting. The reader can identify Mayflies, Lacewings, Crane-flies, Sawflies, Caddis-flies and so on from this book, even though, for example, only 7 of the 193 British Caddis-flies are shown here.

The insects are arranged according to a recent generally accepted classification, with a few exceptions, usually for convenience with the plates. Where there are widely accepted English names, these are used, followed always by the scientific name. In the case of the Lepidoptera, where almost all have well-established English names, these have been used on the plates. But elsewhere the scientific names have been used on the plates, even where English names are given in the text. The descriptions are concerned principally with helping to identify the insect, giving its general distribution, and describing its behaviour which, with many insects, makes fascinating reading. There is little anatomical description, though the chapters at the end of the book give more technical details of structure where these are points of importance in classification.

The pictures, which have been drawn by experienced entomological artists, show, wherever possible, the insect in its natural position on its normal foodplant or other natural setting. Both male and female are usually shown where these differ, and in many cases the larva and pupa. Scale is clearly marked on each page, and where more than one scale on a page was unavoidable, any insect shown at a different scale from the rest of the page is put in a circle.

Readers who have started to use this book may discover the excitement and fascination of finding eggs or larvae of an insect, providing them with their proper food and conditions for development, and then watching the complete stages of metamorphosis, until the adults emerge.

BRISTLE-TAILS, PROTURA,
AND SPRINGTAILS

This page has examples of four different orders of Apterygota, wingless insects, which are mostly very small. These are the Thysanura or Three-pronged Bristle-tails (Nos. 1–2); the Diplura or Two-pronged Bristle-tails (No. 3); the Protura (No. 4); and the Collembola or Springtails (Nos. 5–8). Most of them live in humid situations such as in the soil, leaf litter, or rotting wood, or under logs and stones.

1-2 Three-pronged Bristle-tails (Thysanura). Only 9 species are so far known to live in the British Isles, and these are difficult to distinguish from each other; also not much is known about their life-cycles. The females of some species lay their eggs in batches in late summer and autumn, and the nymphs hatch in the following spring, taking about a year to mature. Most live in various open-air habitats such as heaths, moors, woods, and rocks along the shore, but the two best-known species, the Silverfish (*Lepisma saccharina*) (1), and the Firebrat (*Thermobia domestica*) (2), are common in buildings throughout the British Isles. The Silverfish, which is about ¼ inch long, may often be seen running about kitchens at night, feeding on starchy food scraps, and moving off with great rapidity when discovered. The Firebrat, which is often a little larger, with longer antennae and tail-feelers, also feeds on starchy food scraps, but requires much warmer surroundings than the Silverfish. The female lays her eggs at any time of the year; they hatch in about a fortnight, and the nymphs become adult in about three months. Firebrats have a shorter life-cycle than most Thysanura.

3 Two-pronged Bristle-tail (*Campodea staphylinus*). The 12 British species of the order Diplura are difficult to distinguish from each other, identification depending upon the arrangement of hairs on the back of the thorax and abdomen. They live in humid situations, such as in the soil, among fallen leaves, or under stones; otherwise little is known about their life histories. This species, less than ¼ inch long, is common throughout the British Isles.

4 Acerentomon affine. The tiny, white, eyeless insects of the order Protura have no antennae, but use their forelegs as feelers instead, holding them before their heads. When the larvae first emerge from the eggs, they have only nine abdominal segments, but with each moult they gradually develop the twelve segments of the adult. Like the Two-pronged Bristle-tails, they live in damp places, such as soil in pastures, under stones, and amongst the leaf litter. They are believed to feed on decomposing plant and animal matter. There are 12 British species, and they are most active in the summer. This species, *A. affine*, is probably common in many parts of the British Isles.

5-8 Springtails (*Collembola*). There are about 300 British species of these abundant soil insects, the majority of which have forked springing organs towards the end of the abdomen whereby they project themselves forward. As some species lay their eggs throughout the year, adults and nymphs may be found at any time. They are often so abundant that when Professor E. B. Ford of Oxford investigated an acre of English meadowland to a depth of 9 inches, he estimated that there were altogether about 230 million Springtails of various species within this area. Some feed on decomposing plants and animals and thus may play an important role in improving the soil. Certain species, such as the green *Sminthurus viridis* (5), are pests of plants in Australia and South Africa; but though very common in grass and clover from May to September in many parts of the British Isles, they are not apparently a nuisance here. Some are attacked by red mites. The black *Anurida maritima* (6) is common on sandy beaches and around rock pools on many British coasts, where it is apparently regularly submerged at high tide. Unlike most Springtails, it has no springing organ and can only walk. *Podura aquatica* (7) is locally common on the surface of stagnant water in some districts of the British Isles, including Ireland. *Tomocerus longicornis* (8) is so named because of its very long antennae, which are longer than its body. It is common on logs and in leaf litter on the soil in most places, particularly woodlands, throughout the year.

MUCH ENLARGED (*Nos.* 1, 2 × 6; *No.* 3 × 10; *No.* 4 × 22; *Nos.* 5, 6, 8 × 15; *No.* 7 × 12)

1 LEPISMA SACCHARINA	2 THERMOBIA DOMESTICA	4 ACERENTOMON AFFINE
3 CAMPODEA STAPHYLINUS	6 ANURIDA MARITIMA	5 SMINTHURUS VIRIDIS
7 PODURA AQUATICA	8 TOMOCERUS LONGICORNIS	

1

For a general note on Dragonflies see p. 4. Damselflies (Zygoptera) are mostly smaller than Dragonflies and have a feebler and more fluttering flight. They normally rest with their wings together over their backs. They feed, both as nymphs and adults, on any living creature they can capture, and the nymphs live underwater until they are ready to emerge. Their breathing gills at the tip of the abdomen are much more noticeable than those of Dragonflies.

1 **Demoiselle Agrion** (*Agrion virgo*). A comparatively large Damselfly, of which the male has iridescent wings, which flash deep-blue, purple, or green in the sunshine. The female has smoky-brown wings. The males of both this and *A. splendens* (2) open and close their wings rhythmically when courting a female. They are to be seen from late May to August, flying rather like butterflies, on sunny days, by tree-flanked, swift-flowing rivers and streams, and sometimes lakes, provided they have sandy or gravelly bottoms. The Demoiselle is locally common in the British Isles, especially in the south-west. The brown or greenish nymphs (1c) dwell among the roots of aquatic plants, hibernating during the winter at the bottom of their habitat and usually taking two years to complete their growth.

2 **Banded Agrion** (*Agrion splendens*). The iridescent coloured patches on the male's wings give the insect a banded appearance. The adults, particularly the males, flutter like butterflies in the sunshine from late May to August over slow-flowing streams, rivers, and lakes, especially those with muddy bottoms, in many districts of the British Isles, except Scotland and much of northern England and Wales. It is common in southern England and much of Ireland. The nymphs, which live among the roots of water plants on the mud bottom, vary in colour from chestnut brown to green, and usually take two years to mature.

3 **Green Lestes** (*Lestes sponsa*). This small, metallic blue-green or green-bodied Damselfly is much like its close relative, the very rare *L. dryas*. The males of the latter, however, have bright instead of dull blue eyes, and the females are much smaller and slimmer. The female Green Lestes is usually a duller, more bronze-green with a thicker body than the male. The adults fly from early July to September, and are locally common in most parts of the British Isles, especially the south. They rarely leave the cover of the luxuriant vegetation bordering the ponds, ditches, and other stagnant waters. The female often inserts her eggs in the stems of plants well below the water surface, the male holding on to her while she is doing so. The slim, green or brown nymphs hatch the following spring and live among the water plants near the surface, completing their growth in two to five months.

4 **Large Red Damselfly** (*Pyrrhosoma nymphula*). A very common Damselfly almost everywhere in the British Isles and much larger than the only other red British Damselfly, the very local *Ceriagrion tenellum*, which has red instead of black legs. The male, but not the female, has red eyes. The adults of both sexes are generally yellow at first, growing gradually redder. They usually fly from early May to August in the neighbourhood of either still or flowing water, such as lakes and rivers, at the bottom of which the short, dark-brown nymphs live and feed among the water weeds. These usually take two years to complete their development.

5 **Common Ischnura** (*Ischnura elegans*). One of the commonest British Damselflies, being found almost everywhere. It is difficult to distinguish from the rare *I. pumilio*, from which it can be told only by the arrangement of blue on the abdomen and by the diamond- rather than square-shaped, coloured membrane at the tip of each wing. The females are rather variable in their colour patterns. The adults fly from late May to early September near slow-moving or stagnant water, over the water weeds of which they may often be seen hovering in the sunshine. The male, unlike many Damselflies, does not normally accompany the female while she is egg-laying. The green or brown nymphs are small and slim, and usually take two years to attain maturity.

6 **Common Blue Damselfly** (*Enallagma cyathigerum*). Almost, if not as common as the Common Ischnura, this pretty Damselfly is found throughout the British Isles, often far from water. The adults, which fly from mid-May to early September, often swarm around lakes, rivers, and smaller waters, particularly in marshes, even if the water is brackish. There are several other blue British Damselflies, but this species can be recognized by the entirely blue 8th and 9th segments of the male's body. It differs from species of the genus *Coenagrion* in possessing only a single, short, black line on the side of its thorax, instead of two. The grey-green and black females have an easily-seen spine beneath the 8th abdominal segment. They often insert their eggs in plant-stems below the surface, completely submerging themselves to do so; they either pull the attendant males down with them, or leave them waiting above. The slim nymphs (6c), which hide among the aquatic plants, vary considerably in colour, which adapts to match their surroundings. They take about a year to reach maturity.

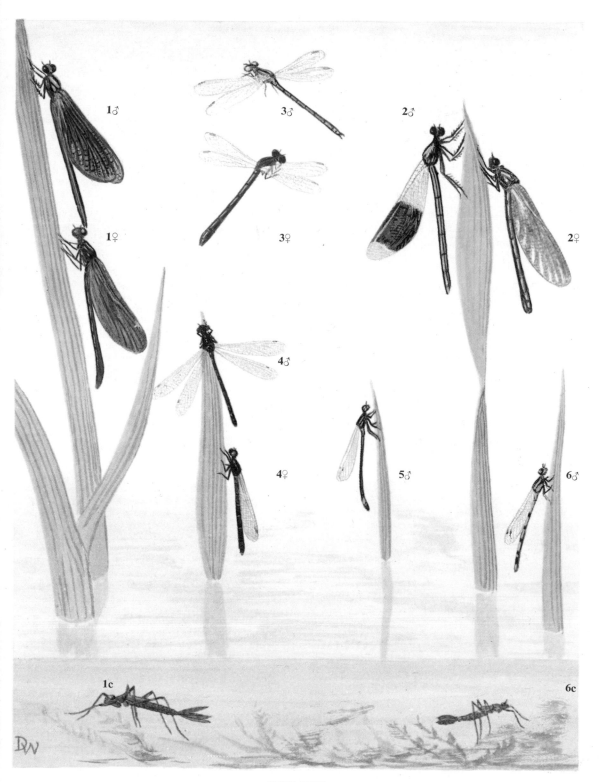

LIFE SIZE

1 AGRION VIRGO + NYMPH (1c)   3 LESTES SPONSA   2 AGRION SPLENDENS
4 PYRRHOSOMA NYMPHULA   5 ISCHNURA ELEGANS   6 ENALLAGMA CYATHIGERUM + NYMPH (6c)

3

There are two sub-orders of Dragonflies: the Zygoptera or Damselflies (*see* p. 2) and the Anisoptera, the typical Dragonflies (*see* p. 195). They all lay their small, yellowish eggs directly into the water or inserted into water plants, floating debris, or mud. The males in many species help by holding the females while they are egg-laying and then pulling them up into the air again.

The larvae or nymphs take from one to five years to complete their development, according to species. There is no pupa stage. When nearly mature, they climb up a plant stem out of the water, shed the last larval skin, expand and dry their wings, and fly off as perfect adults. Both adults and nymphs are predatory, eating whatever they can capture. The nymphs have pincer-like extensions of the lower lip, called the 'mask', which they can shoot out to catch their prey. They have breathing gills which, with most species, are at the tip of the abdomen.

Dragonflies are swift and agile fliers — hence the name 'hawker', and they have very large compound eyes. Unlike Damselflies, they usually rest with their wings spread out. The adults usually live about a month.

1 Golden-ringed Dragonfly (*Cordulegaster boltoni*). A very large, black-and-yellow-banded Dragonfly, the only British member of its family. The sexes are alike, except for the female's hollow, black ovipositor-sheath, which projects beyond her abdomen and which she uses to push her eggs into the mud or ground beneath the water. While laying her eggs, the female hovers in undulating flight. The hairy, dark-brown nymphs hide in the mud under the water, with their eyes and the tips of their abdomens showing, and take two to five years to reach maturity. The adults fly very powerfully by day and even after dark from June to September, hawking up and down moorland and mountain streams and rivers in search of their insect prey. They are locally common in such habitats, chiefly in northern and western Britain; but they are rare elsewhere and unknown in Ireland.

2 Common Aeshna (*Aeshna juncea*). This very large Hawker Dragonfly is the most widespread of the Aeshnidae, a family to which the remaining species on this page belong. They mostly have long, slim bodies and fast, powerful flight. This species is like *A. cyanea* and the rare *A. mixta*, but the mature male is much bluer than the former and much larger than the latter, and has golden-yellow leading edges to the wings. Females and immature males are normally green or yellow, without the adult male's bright blue eyes. The Common Aeshna is the only common Hawker Dragonfly in the north and west, including Ireland. It prefers acid water and is therefore commonest on heaths and moors; consequently *A. cyanea* (the Southern Aeshna) is often the commoner in southern England. The adults fly by day from July to October, often far from water, and may be seen patrolling, for example, a woodland glade, even at dusk. The dark-brown nymphs live among the water weeds, usually taking at least two years to become full grown.

3 Brown Aeshna (*Aeshna grandis*). A very large Hawker Dragonfly, larger than the only other tawny-brown member of the family, the Norfolk Aeshna (*A.*

isosceles). The male has blue eyes, blue spots on the base of the abdomen, and yellowish wings when mature. The female has yellow-brown eyes. The Norfolk Aeshna, which is more or less confined to the Norfolk Broads, has green eyes, a yellow triangle on the base of the abdomen, and colourless wings. The Brown Aeshna is widespread and common over much of England and Ireland, but is little found in the north. Migrants often reach Britain from the Continent. It is on the wing from July to early October, frequently flying until dark, and often well away from water and even in towns. It is most often seen around quiet ponds, lakes, rivers, and disused canals, in which the female deposits her eggs in plants or floating wood. The dark-brown or green nymphs take from two to five years to mature.

4 Emperor Dragonfly (*Anax imperator*). With a wingspan of up to 4 inches and body length of nearly 3½ inches, this is probably the largest British Hawker Dragonfly. It differs from other Hawkers in the shape of the hind-wings; the inner margins being rounded instead of angled, and in having no bands on the thorax. Both sexes have a black, stripe-like pattern along the top of the abdomen and are without the chequered appearance of other Hawkers. The adults may be seen patrolling and hunting in the sunshine and towards nightfall from June to early September around ponds and lakes in many parts of England; but they are common only in the south-east and are never found in Scotland. Like other large Hawker Dragonflies, they prey upon smaller Dragonflies and many other insects, which they can out-fly and overpower. The smooth, light-brown, green, or yellowish nymphs (4a) are extremely active and aggressive, hunting among the water plants not far below the surface of the water. They take two years to mature and, when moulting, they change their colour to harmonize with their environment. The younger nymphs, like those of most of the Aeshnidae, often have black and white bands which break up their outline and help them to hide from the cannibalistic older nymphs.

LIFE SIZE

1 CORDULEGASTER BOLTONI   2 AESHNA JUNCEA
3 AESHNA GRANDIS   4 ANAX IMPERATOR + NYMPH

DARTER DRAGONFLIES

For a general note on Dragonflies see p. 4. All the species on this page, except the Downy Emerald (No. 1), are known as Darter Dragonflies (family Libellulidae), because of their habit of repeatedly darting out on a brief flight from a favourite perch and then returning to it again.

1 **Downy Emerald** (*Cordulia aenea*). The least uncommon of the four British species of the family Corduliidae — medium-sized, metallic-green Dragonflies — is so called because of the dense yellow hair on the thorax. It is less bright green than the other species, and normally has orange-yellow patches at the base of all four wings. It flies up and down, low over the surface of lakes, ponds, canals, and slow-flowing rivers and streams from May to late July, being locally common in south-eastern England, but rare or absent elsewhere, except for a few scattered localities in Scotland and Co. Kerry, Ireland. The females hover alone, without the males, when egg-laying, dipping their abdomens into the water. The long-legged nymphs (1a) usually take two years to mature, during which time they live on the muddy bottom of their habitat.

2 **Four-spotted Libellula** (*Libellula quadrimaculata*). The largest of the family, these Dragonflies often move in massive numbers in northern Europe on migrations, sometimes so as to darken the sky. The most recent large movement occurred in the North Sea area in June, 1963. The sexes are alike, both having two spots on each wing and black patches on the base of the hindwings. Both this and *L. depressa* (No. 3) have much thicker bodies than most Dragonflies. It is widespread, though rare in some parts of Scotland and northern England, and in good migration years it may become extremely numerous. The adults are on the wing from May to late August, usually hunting their insect prey near any kind of stagnant or slow-moving water; but, especially when on migration, they may be seen well away from water and even flying swiftly along city streets. The females lay their eggs in the same way as Downy Emeralds do, and the dark-brown nymphs are mud-dwellers, usually half-burying themselves in mud and debris. They take more than two years to mature.

3 **Broad-bodied Libellula** (*Libellula depressa*). Another migrant Dragonfly. Both sexes have very broad, flattened abdomens — the mature adult males being blue and the young males and females being brown. They fly by day from May to August in much the same habitats as *L. quadrimaculata* (No. 2), but rarely breed north of Lancashire and Yorkshire, or in Ireland. They are very common in southern England, frequently breeding in garden pools. The nymphs are similar to those of *L. quadrimaculata*, but hairier and more mottled; they live in the same situations.

4 **Common Sympetrum** (*Sympetrum striolatum*). This and the next two species are medium-sized Darter Dragonflies belonging to a genus the members of which are not always easy to distinguish from each other. They differ from *Libellula* in having slender abdomens, which in some species (*see* Nos. 5 and 6) are club-shaped. The different species have different vein patterns and colours, and some species have extensive orange-yellow or red patches on the base of the wings. The Common Sympetrum is usually abundant in the south, especially when reinforced by larger and paler migrants from the Continent, and is found in most parts of the British Isles, although local or absent in parts of Scotland. They behave like other Darter Dragonflies, and are often found far from water, sometimes swarming in grassy fields or along paths and roads, from late June to late October. They frequently migrate in great swarms. When the female is egg-laying, the pair usually fly slowly 'in tandem' over the surface of the water, dipping repeatedly so that the eggs on the female's abdomen are washed off and sink below (*see* p. 194). The spiny nymphs (4a), which vary from light to dark brown according to their surroundings, dwell among the water weeds, preferably in fresh water, and complete their growth in a year.

5 **Ruddy Sympetrum** (*Sympetrum sanguineum*). The deep crimson-red colour and club shape of the adult male's abdomen distinguish it from others of the genus (*see* No. 4), but the young males and females are more difficult to identify. The Ruddy Sympetrum, which is on the wing from the end of June to September, is also a well-known immigrant, large numbers sometimes arriving in the British Isles from the Continent, when it will appear in districts where it does not normally breed. The resident British ones breed only in lakes, ponds, gravel-pits, and dykes in south and east England where great reed-mace and horsetails grow; there, they are locally common. They are rare in Ireland and Wales and absent from Scotland. They behave much like the Common Sympetrum but have a more indecisive, fluttering flight and a stricter adherence to water. The dark-brown nymphs hide among the submerged roots of great reed-mace and horsetails and mature within a year.

6 **Black Sympetrum** (*Sympetrum danae*). The mature males, with their black, club-shaped bodies and black legs, are distinct from most other members of the family. The females and young males are generally darker than other Sympetrums, and both males and females have a dark, triangular mark on the top of the thorax, just behind the head. They fly from mid-July to late October and are common, especially in Scotland, wherever there are acid bogs, marshes, pools, and lakes in which to breed. They sometimes migrate in swarms. Their habits and life-cycle are similar to other Sympetrums.

LIFE SIZE

1 Cordulia aenea + nymph   2 Libellula quadrimaculata
3 Libellula depressa   4 Sympetrum striolatum + nymph
5 Sympetrum sanguineum   6 Sympetrum danae

1-3 Mayflies (*Ephemeroptera*). Most British members of this order of insects are small and all are usually found near water. They can be easily distinguished from Stoneflies because, when resting, they hold their wings vertically over their bodies. Most have four wings, the hindwings being small; in some genera there are no hindwings. The eggs are laid in or on fresh water, and the larvae (nymphs) live in the water, and most are vegetarian. They vary greatly in the time they take to mature; this may be up to three years or short enough to allow three generations in a year. When ready to emerge, they come to the surface of the water, cast the final larval skin, and fly weakly. But, unlike any other insect, they still have to undergo a further moult before they are complete and able to fly properly and reproduce. The adults take no food and live at the most three or four days. Swarms of males perform a mating dance, and when one encounters a female, he flies beneath her, grasps her with his long forelegs and pair of short claspers, and they mate. The female lays her eggs in the water, dipping her abdomen below the surface, and soon after, dies, often on the water.

Fishermen use Mayflies or imitation Mayflies for freshwater fishing and have given them names. They call the immature adults 'duns' and the complete adults 'spinners' and have special names for certain species.

Ephemera danica (1) called by fishermen 'Green Drake', is the best-known and largest British Mayfly. It is on the wing from mid-April to September, often in vast numbers, when fish, especially trout, eat quantities of them. The nymphs (1a) burrow in the muddy bottoms of rivers and lakes. The widespread and often common little *Cloëon dipterum* (2) breeds in still waters, even in water-butts and cisterns. The adults have only two transparent wings, those of the females having yellowish front edges, and there are two instead of three tails. They have olive-brown bodies and are called 'Pond Olives' by fishermen. The little *Caenis rivulorum* (3) is more local than the others and belongs to small rivers and streams. The adults emerge, often in large numbers, between June and September, and are also eagerly eaten by trout.

4-7 Stoneflies (*Plecoptera*). These, like Mayflies, are found near water, but, when resting, tend to sit with their wings either folded flat or rolled tightly around their bodies. They are also used as bait by fishermen, who have given them names. They often call the nymphs 'creepers' because they tend to creep among the stones on the water bed. Some are carnivorous, but most are vegetarian. Most take a year to complete their development. The adults take some food, either liquid or solid, and they live for two or three weeks.

Perla bipunctata (4), one of the two largest British Stoneflies, is a robust insect with a wingspan of 2 inches or more. It is common over the British Isles, except for the east and south-east. The black-and-yellow, carnivorous nymphs creep among stones on the beds of fast-flowing rivers, and the adults, when they appear in May or June, hide in rock crevices at the water's edge, not flying a great deal. Its close relative *Dinocras cephalotes* (5) closely resembles it but is without the yellow just behind the head. Its habits and distribution are similar, but the nymphs prefer rivers with firmer bottoms.

Capnia vidua (6) is one of the smallest British Stoneflies. The males have such reduced wings that they are useless for flying. They are scarce, frequenting small, stony streams in a few localities in the Midlands and North. *Leuctra fusca* (7) is a common Stonefly of rivers, streams, and lakes, and flies from August to October. Anglers call it the 'Needle-fly' because, when it rests, it rolls its dark-brown wings tightly around its body. The nymphs feed on aquatic plants.

8-9 Earwigs (*Dermaptera*). There are only seven British species of Earwigs, one of which, the yellow-brown Giant or Tawny Earwig (*Labidura riparia*), is extremely rare, or possibly extinct. Those Earwigs which fly have transparent and very thin hindwings, which, when not in use, are folded up in a remarkable way and are mainly hidden by the forewings, which are no more than short wing-cases, like those of beetles. The forceps at the end of the abdomen, usually more curved in the male than in the female, are probably mainly for attack or defence, but also help in folding their wings. The idea that Earwigs get into human ears has no more foundation than that Earwigs will go into any dark crevice during the daytime. The idea is widespread, however, for in German they are called *Ohrwurm* (ear-worm) and in French *Perce-oreille*.

The only Earwig to be fully studied is the Common Earwig (*Forficula auricularia*) (8), which, though possessing wings, rarely flies. This species, as probably the others, is an omnivorous feeder, and when seen in flower-heads or fruit, may be pursuing insect prey rather than feeding on the plants. A male mates with only one female. After mating, they hibernate in underground cells. In very early spring, the female lays a batch of white eggs in the cell, and when the male leaves the cell, she remains to guard the eggs. They hatch into tiny Earwigs, like the adults, and the mother protects them until they can fend for themselves. In late June or July they are mature enough to mate. The females may have more than one brood in a season, after which they die.

The Lesser Earwig (*Labia minor*) (9) is common in rough herbage such as nettle beds in many parts of the British Isles. The adults fly actively in August and September and sometimes come to artificial light at night.

TWICE LIFE SIZE

1 Ephemera danica + nymph   2 Cloëon dipterum   3 Caenis rivulorum
4 Perla bipunctata + nymph   6 Capnia vidua   5 Dinocras cephalotes
7 Leuctra fusca   9 Labia minor   8 Forficula auricularia

Only three of the nine British species of Cockroaches (Dictyoptera) are truly native. They look rather like beetles, but are no relation. The eggs, laid in leathery, purse-like cases called oothecae, hatch into worm-like grubs, which moult their skins almost immediately, and then resemble the adults increasingly with each subsequent moult (*see* p. 195).

The difference between Crickets and Mole-crickets and Bush-crickets, all Orthoptera, is described on p. 12. Male Crickets stridulate by rubbing a toothed rib or 'file' in the right forewing against the hind margin of the left forewing — the reverse of the Bush-cricket method. As with Bush-crickets, their 'ears' are situated on their front legs, but their heads are more rounded. Female Crickets have long, needle-like ovipositors, and Mole-crickets have forelegs specially adapted for digging. Both are omnivorous.

1 **Common Cockroach** (*Blatta orientalis*). Generally known as the 'Blackbeetle' of old kitchens, bake-houses, and such places, this large pest of dirty places was accidentally introduced into the British Isles about the 16th century, and is now widespread and common, though rarely found out-of-doors, except perhaps on a rubbish dump during the summer. Its wings are so little developed that it cannot fly. It hides during the daytime and emerges at night to feed on almost anything remotely edible, including even leather, paper, and wool. The female's egg-case is white at first, but darkens until it is almost black. She carries it round for a few days and then deposits it in a suitable place, such as a warm crevice, and cements it firmly in position. Each case usually contains fourteen to sixteen eggs arranged in two rows, and these usually hatch two or three months later. The nymphs normally take nearly a year to mature. All stages of the Common Cockroach may be found throughout the year.

2 **American Cockroach** (*Periplaneta americana*). This species, larger and paler than the Common Cockroach and with fully-developed wings, was introduced possibly from America, though it is not a true native of America. It is most common in warm buildings, such as breweries and warehouses, chiefly in coastal towns.

3-4 **Cockroaches** (genus *Ectobius*). The Tawny Cockroach (*E. pallidus*) (4) is the only British species in which both sexes can fly. Only the males of the other two native species, the Dusky Cockroach (*E. lapponicus*) (3) and the Lesser Cockroach (*E. panzeri*), can fly. All these species are mainly restricted to southern England. The Tawny Cockroach, which has also been found in South Wales, frequents low vegetation in woods, and on heathland, commons, and chalk-hills. The rather broad female deposits her dark-brown egg-cases on the ground between June and September. These hatch next spring, and the nymphs hibernate the following winter when nearly full grown. They awake in May, and mature and mate during the summer. On sunny days they run and fly very actively.

5 **Wood-cricket** (*Nemobius sylvestris*). This smallest British cricket has flap-like forewings and no hind-wings. The females lay their eggs singly in the soil in late summer and autumn; the nymphs hatch the following June, hibernate, and mature in the July after that. The majority die during the next winter, but a few survive into the next year. They are most active by day, running and jumping in warm weather with a curiously jerky action, and the males' song, rather like a distant Nightjar, may be heard also on warm nights. They inhabit leaf-litter in woodland borders and clearings and are locally common in parts of south-west England, such as the New Forest, and now also in Surrey.

6 **Field-cricket** (*Gryllus campestris*). This rare, flightless Cricket is found only on a few sunny banks and fields in southern England, where there may be largish colonies. The males' shrill chirp can be heard continuously from May to August from near the burrows which they make in the soil, and near which the females lay hundreds of eggs during the late spring and summer. The nymphs live above ground until the autumn, when they burrow into the soil and hibernate; they awake early next spring, reach maturity by early May, and die off in August.

7 **House-cricket** (*Acheta domesticus*). Though once familiar, the shrill chirp of this insect is now seldom heard because the old buildings where it usually lives are kept too clean. It originally came from North Africa and south-west Asia and reached Britain about the same time as the Common Cockroach. It is still found in many parts of Britain, though rarely out-of-doors, except on rubbish dumps, where it is often numerous. Adults may be found throughout the year and fly well in warm weather, when the males 'sing', chiefly in the evening and at night. A single female lays groups of several hundred eggs in crevices and such places. These normally hatch two three or months later, and the nymphs mature in about six months.

8 **Mole-cricket** (*Gryllotalpa gryllotalpa*). A large but now rather rare insect, to be found in damp meadows and fields in certain river valleys in south Hampshire and Surrey. The adults fly noisily and heavily on warm spring and summer nights, when the males also 'sing' continuously from their burrow entrances. After hibernating, the females lay their numerous eggs in underground nests and guard them until they hatch a few weeks later. The nymphs hibernate underground and do not usually become adult until the following autumn.

ONE-AND-A-HALF TIMES LIFE SIZE

1 BLATTA ORIENTALIS   2 PERIPLANETA AMERICANA + EGG CASE
3 ECTOBIUS LAPPONICUS   4 ECTOBIUS PALLIDUS
6 GRYLLUS CAMPESTRIS   5 NEMOBIUS SYLVESTRIS   7 ACHETA DOMESTICUS
8 GRYLLOTALPA GRYLLOTALPA

BUSH-CRICKETS (*Tettigoniidae*)

The Bush-crickets, or 'Long-horned Grasshoppers' differ from the true Grasshoppers (p. 14) in having long, thread-like antennae instead of short ones; they differ from Crickets and Mole-crickets (p. 10) in having tarsi (feet) divided into four instead of three segments. Their hearing organs are situated on their front legs. The males produce a 'song' by rubbing a toothed rib (the 'file') in the left forewing against the hind margin of the right forewing, where a membrane amplifies the sound. The males have pincer-like appendages (cerci) on the tip of the abdomen, and the females have flattened, sword- or sickle-shaped ovipositors. Adults and nymphs eat both vegetable and animal food, including insects. The females lay their eggs singly either in or on plants, or in the ground, and these hatch the following spring into worm-like larvae which, on reaching the surface, moult. They then become more and more like the adults with each successive moult.

1 **Oak Bush-cricket** (*Meconema thalassinum*). This is the only British Bush-cricket to live in trees, especially oak. It is common in most English and Welsh woodlands, particularly in the south, but is not found in Scotland, and in Ireland only in County Limerick. The adults hide among foliage by day, becoming active after dark, normally from August to November. They are capable of weak flight and may come indoors, attracted by lights. The females usually lay their eggs in bark crevices, and these hatch the following June. The males have no obvious sound-producing structures, but they produce short bursts of 'song' by drumming one hind leg very rapidly on a leaf, and a feeble sound by rubbing their forewings together.

2 **Great Green Bush-cricket** (*Tettigonia viridissima*). The loud, almost continuous, penetrating 'song' of the males is a feature of warm summer nights in parts of the English countryside, especially near the south coast and along the south coast of Wales. It becomes less common further north, and is not heard in Scotland and Ireland. The insects lurk in the thick vegetation of hedgerows, nettle-beds, bramble thickets, and reed-beds, and if discovered, try to escape by running, with occasional hops or short flights. They are most active, hunting other insects, from mid-day to around mid-night from late July to early October. The eggs, which are deposited in the soil, overwinter and hatch in late spring, and the nymphs mature in late July or August.

3 **Dark Bush-cricket** (*Pholidoptera griseoaptera*). On warm afternoons and nights, from early August to early November, the males of this medium-sized Bush-cricket chirp in chorus from the hedgerows, bramble clumps, and nettle-beds where they hide. They are generally common throughout southern England and South Wales, becoming more local further north, and not found north of Yorkshire or in Ireland. The females, which are not unlike large, dark spiders, have only vestigial wings, while the smaller, darker males have merely flaps, consisting only of the sound-producing organs; their normal 'song' is a brief, metallic chirp repeated intermittently for long periods. The females lay their eggs in bark and other crevices; the nymphs hatch the following spring and mature by August.

4 **Bog Bush-cricket** (*Metrioptera brachyptera*). This is a medium-sized species with two colour forms, a green and a brown, the former being the commoner. It differs in small points of colouring from the very local Roesel's Bush-cricket (*M. roeselii*): for instance, its underside is bright green instead of yellow. Very occasionally, individuals are found with fully-developed wings. It is common over much of England and Wales on heath and moorland bogs, and in nearby damp vegetation, and is active by day from July to October. The eggs, laid in plant stems, hatch the following spring. The 'song' consists of repeated short, shrill chirps, five or six per second, lasting for minutes on end.

5 **Short-winged Conehead** (*Conocephalus dorsalis*). This Bush-cricket, which moves like quick-silver through the vegetation of its marshy habitat, can be distinguished from its much rarer relative, the Long-winged Conehead (*C. discolor*) by the curved instead of almost straight ovipositor of the female, and by the short, not fully-developed wings. (Very rarely individuals with fully-developed wings occur.) It is found mainly in the southern and eastern seaboard counties of England and South Wales. Adults usually occur from August to early October, being active only by day. The females lay their eggs singly in the stems of rushes and sedges, and the nymphs appear in May or June. On warm, sunny days the male produces a quiet, high-pitched, reeling 'song', which may continue without a break for up to 2 minutes.

6 **Speckled Bush-cricket** (*Leptophyes punctatissima*). A plump Bush-cricket, common in wood borders, hedgerows, bramble thickets, nettle-beds, and similar situations — even in gardens — in many districts of Wales and of England as far north as south-west Yorkshire. It has also been found in Wigtownshire, Scotland, and in the Isle of Man and south Ireland. The female's wings are vestigial; the male's are small and flap-like, the sound-producing organs being capable only of a very weak chirp. The adults are active on warm, sunny days from August until October The females lay their curious, flattened eggs in plant stems and bark crevices, and these hatch the following May or June.

1♀   3♂   2♀   6♀   5♀   4♀   4♂   5♂

DW

ONE-AND-A-HALF TIMES LIFE SIZE

1 MECONEMA THALASSINUM   3 PHOLIDOPTERA GRISEOAPTERA

2 TETTIGONIA VIRIDISSIMA

4 METRIOPTERA BRACHYPTERA   6 LEPTOPHYES PUNCTATISSIMA   5 CONOCEPHALUS DORSALIS

13

GRASSHOPPERS (*Acrididae*)

The true or Short-horned Grasshoppers differ from Bush-crickets (p. 12) and Crickets (p. 10) in possessing short, thickish antennae instead of long, thread-like ones. Their hearing organs are situated on each side of the abdomen, near the thorax. The males of all British species, and the females of some, 'sing' by rubbing the inside of the hind-leg, studded with tiny pegs, against thickened veins on the forewings, causing them to vibrate (*see* p. 195). The British species are active by day, and all, except the Meadow Grasshopper, have well-developed wings and can fly. But they move mainly by hopping, and have long hind-legs with very powerful thighs (femurs), which enable them to make long jumps. The males are generally smaller than the females, and the tips of their abdomens are boat-shaped, whereas those of the females are somewhat rounded. The eggs, laid in batches, are enclosed in strong egg-pods and pass the winter in this stage. The nymphs hatch in the spring; both they and the adults are entirely vegetarian, feeding chiefly on grass.

1 **Large Marsh Grasshopper** (*Stethophyma grossum*). The largest British true Grasshopper is found commonly only in wet bogs and fens, especially where bog myrtle and bog asphodel grow. Such localities are sparsely scattered over southern England and western Ireland, the best known being in the New Forest. This Grasshopper is easily recognized by its bright hind-legs. The males are mainly greenish-yellow, and the larger, bulkier females are olive-green or brown, though sometimes rose-purple. The male's 'song' is unmistakable — a short, slow ticking, produced by flicking the hind-legs several times upwards and backwards on to the folded forewings. On hot, sunny days, mainly in August and September, they make prodigious leaps and flights, and are very hard to catch. The egg-pods, containing a dozen or so eggs, are deposited beneath tufts of grass, and the eggs hatch in late spring.

2 **Stripe-winged Grasshopper** (*Stenobothrus lineatus*). A locally common insect in southern England only, in dry places on sandy or chalky soils, especially where the grass is short. Both sexes are mainly green, usually with a white, curved streak on each forewing; the females often have a white line in addition. They deposit egg-pods with up to eight eggs at the base of grass-tufts, where they overwinter and normally hatch in May. The nymphs reach maturity from late July. The males have a distinct courtship song and also a rather quiet but continuous song, with a metallic scouring quality, which can be heard until September.

3 **Common Green Grasshopper** (*Omocestus viridulus*). Found almost everywhere in the British Isles in a wide range of grassy habitats; this is the only species in many high moorland areas. It is usually emerald-green or olive-brown, but with a fair amount of variation. It resembles the rare Woodland Grasshopper (*O. rufipes*), but has no orange-red colour on the underside of the abdomen, and the penetrating, fast, continuous

'ticking' song of the male distinguishes it from the Stripe-winged Grasshopper (No. 2). The adults jump and fly actively in the sunshine from late June to early October. The egg-pods are laid under grass-tufts and hatch in the following spring.

4 **Common Field Grasshopper** (*Chorthippus brunneus*). One of the largest and commonest British Grasshoppers, although not as widespread in northern Britain, or on high ground as the last species. It has a wide range of habitats — fields, dunes, open woodlands, city waste ground — but it prefers dry situations. The adults are extremely variable in colour, but are usually some shade of brown; distinctive features are the sharply indented side keels on the thorax and the very hairy underside, and the long forewings (reaching beyond the hind knees when at rest). They are very active and strong fliers in hot sunshine. The male's 'song', often in chorus since this Grasshopper is very gregarious, consists of a series of short chirps, and is to be heard from late June to mid-November, or even later. Up to fourteen eggs are contained in the egg-pods, which are deposited in the ground and hatch the following spring.

5 **Meadow Grasshopper** (*Chorthippus parallelus*). This species, very common in almost all parts of the British Isles, except Ireland, the Isle of Man, and many islands off Scotland, likes grassy places, provided they are not on high hills or too dry. The vestigial hindwings, which prevent both sexes from flying, differentiate them from other species. Sometimes, however, a fully-winged variety occurs. They vary in colour, but most are green or greenish. Adults occur from the end of June to late October. The males produce a very short, somewhat chattering 'song' on warm days, and occasionally at night, often answering each other, for they are gregarious. The females lay egg-pods in the soil, each containing up to ten eggs which normally hatch in April or May.

ONE-AND-A-HALF TIMES LIFE SIZE

1 STETHOPHYMA GROSSUM   2 STENOBOTHRUS LINEATUS
3 OMOCESTUS VIRIDULUS   4 CHORTHIPPUS BRUNNEUS   5 CHORTHIPPUS PARALLELUS

15

GRASSHOPPERS AND GROUNDHOPPERS

A general note on the true Grasshoppers is given on p. 14.

1 Lesser Marsh Grasshopper (*Chorthippus albomarginatus*). The straight, parallel side-keels of this medium-sized Grasshopper distinguish it from the related Meadow and Common Field Grasshoppers (p. 14). There are several colour varieties, green, brown, and straw-coloured forms being the most usual. The females, which often have a white streak on the forewings, deposit egg-pods (each with up to ten eggs) at the base of grass-tufts. The nymphs hatch the following spring, become adult in late July and August, and usually die off by early October. On warm, sunny days the males produce their normal 'song' — a short series of quiet chirps; their courtship song sounds like a clockwork toy, whirring intermittently for up to 20 seconds each time. Both sexes have fully-developed wings, and are active fliers — especially the males. They haunt damp, low-lying meadows and marshes, particularly near the coast, but they also frequent sand-dunes. They are common in many parts of England as far north as Lancashire and Yorkshire, and occur in a few places in Wales and Ireland.

2 Rufous Grasshopper (*Gomphocerippus rufus*). This species is almost entirely confined to chalk and limestone hills in southern England and south-east Wales, mainly on south-facing slopes. Its distinctive white-tipped antennae are strongly clubbed (rather less so in the larger female) like those of a butterfly. The Mottled Grasshopper (No. 3) is the only other British species with clubbed antennae. The Rufous Grasshopper is usually some shade of brown, although there is a rare reddish-purple variety of the female. Adults may be found from late July to early November; on warm sunny days they jump and fly actively and stridulate. Their normal 'song' is rather quiet and lasts about 5 seconds, but they also produce a sporadic ticking chirp. For courtship, the males move their heads and antennae in distinct patterns, and repeat the normal 'song' several times, ending each time with a sharp click. After mating, the females deposit egg-pods (each containing up to ten eggs) in the soil, and these hatch the following late May or June.

3 Mottled Grasshopper (*Myrmeleotettix maculatus*). A common species, often abundant on dry moors, heaths and commons, sand-dunes, and chalk and limestone hills (usually where the rock is exposed) throughout the British Isles, except in mountainous country. These small, compact Grasshoppers are mottled or speckled, and vary enormously in colour; dominantly brown, black, and green forms are most frequent. Like the Rufous Grasshopper (No. 2), they have clubbed antennae (only slightly clubbed in the larger females). Adults occur from June to October, and feed on grass. The females deposit their egg-pods (containing up to six eggs) in the soil, and the nymphs hatch the following spring. Both sexes jump actively in sunny weather. The males deliver a series of about twenty short chirps in 10 – 15 seconds — soft at first but steadily growing louder.

4-5 Groundhoppers. These members of the family Tetrigidae have a 'hood' stretching backwards from the thorax over the top of the body. They lack ears, and none of the three British species are able to stridulate. The much reduced forewings are partly hidden by the 'hood'; the hindwings are more fully developed. The British species are small, extremely well camouflaged, and active on hot, sunny days, feeding chiefly on algae and mosses on or near the ground. Groundhoppers differ from true Grasshoppers in overwintering as nymphs or adults instead of as eggs, and in laying their clusters of ten to twenty 'horned' eggs, stuck together by a secretion, in spring and summer, usually in the soil. The nymphs hatch within a few weeks.

The Common Groundhopper (*Tetrix undulata*) (4) is a squat little insect, which differs from the Slender Groundhopper (5), in having a shorter 'hood' (not extending beyond the hind knees), with a much more pronounced central ridge on it, and much shorter hindwings. Very rarely, a fully-winged form occurs. This species varies from pale grey to black, but is usually brownish or blackish. It is found almost everywhere in the British Isles in mossy habitats such as heaths, open woodland, moors, and marshes. It cannot fly, but leaps extremely well and also swims. Nymphs and adults are to be found together throughout the year because the females lay eggs over a long period.

The Slender Groundhopper (*Tetrix subulata*) (5) is a locally common species throughout southern England, and is also sometimes found in Glamorganshire and Co. Galway, Ireland. The 'hood' normally extends well beyond the knees of the hind-legs, although in one not uncommon form it reaches only just beyond the hind knees. The hindwings are fully developed and long. The colour varies widely, but is usually mottled black or brown. This Groundhopper prefers marshy habitats where rushes grow and damp pasture with short grass. It jumps and flies actively in hot sunshine, often leaping several yards, and it swims well. Adults are found in spring and early summer, and again from August onwards, but are dormant in winter. The eggs are laid between May and July. The nymphs hatch in early spring and mature in August; the later ones overwinter as nymphs and become adult the next spring.

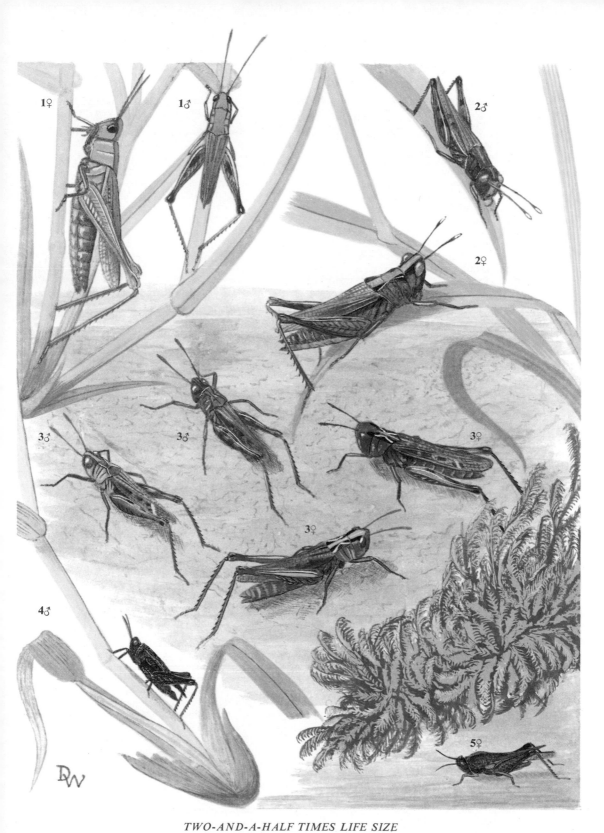

TWO-AND-A-HALF TIMES LIFE SIZE

1 Chorthippus albomarginatus   2 Gomphocerippus rufus
3 Myrmeleotettix maculatus
4 Tetrix undulata   5 Tetrix subulata

There are five different orders of insects shown here: the Psocoptera (Book-lice and Psocids); the Mallophaga (Bird-lice); the Anoplura (Sucking-lice); the Siphonaptera (Fleas) and the Thysanoptera (Thrips). The fleas, close relatives of the Diptera, are put here for convenience.

1-3 Book-lice and Psocids (Psocoptera). Most of this group including those with wings, are found out-of-doors, usually beneath the bark or among the foliage of trees, and on fences or walls. They feed on animal or vegetable matter, and many live in colonies, sometimes forming big swarms. Those we most often see occur in houses, and are usually those without, or with rudimentary, wings, such as *Trogium*. They feed on the paste used in bookbinding, hence the name Book-lice (though they are not Lice at all). Many are found in new houses, feeding upon the minute moulds that grow on still damp plaster and wallpaper paste. The females lay their eggs either singly or in groups, some species protecting them with silken webs or debris. Except for being wingless, the nymphs resemble the adults. According to species and conditions, from one to twelve generations may be produced in a year.

The Common Book-louse (*Trogium pulsatorium*) (1) has wings reduced to mere pads. It is found all over the British Isles in books, furniture, insect collections, and stored foodstuffs, especially where there is damp, when it may produce several broods a year. *Metylophorus nebulosus* (2), one of the largest British Psocids, has four wings and extremely long antennae. It is widespread, especially in conifers and fruit trees such as apple and plum. It probably produces only one generation a year — between mid-July and the end of September — and overwinters as eggs covered with debris. *Lachesilla pedicularia* (3) is a small, winged, widespread species, difficult to distinguish from its relatives, and most commonly seen in the autumn. It is occasionally a pest in stored food and swarms in vast numbers out-of-doors — swarms have been seen in the Thames Estuary and on the Belgian coast.

4-5 Bird-lice or Biting-lice (Mallophaga). The 500 and more British species of these small, flattened, wingless insects are mostly bird parasites, although some infest mammals. Most birds carry them and probably take dust baths to rid themselves of them. The females attach their eggs (nits) to feathers or hair, and both nymphs and adults feed with their biting mouthparts on scurf, hair, or feathers. Some species slso take the blood of their hosts. The Chicken Louse (*Menopon gallinae*) (4) is found commonly on poultry, and the Pigeon Louse (*Columbicola columbae*) (5) lives among the feathers of doves and pigeons.

6 Sucking-lice (Anoplura). These are parasitic on mammals only, sucking their blood. Each species attaches itself to one particular mammal, man being parasitized by the Human Louse (*Pediculus humanus*) (6) which normally attacks no other species. The two varieties, *corporis*, confined to the body, and *capitis*, mainly confined to the head, can interbreed. The females of both varieties cement their 'nits' singly to the hairs or clothing of the host. The nymphs hatch in a week or two, suck blood like the adults, and mature in 8 to 16 days. Human Lice are found all over the world where people live in crowded, unhygienic conditions, particularly among children in crowded towns. They cause much irritation and spread serious diseases such as typhus and louse-borne relapsing fever.

7 Fleas (Siphonaptera). The 47 British species of these blood-sucking insects, are mostly parasites on a particular animal — domestic fowls, hedgehogs, cats, and so on. Man is usually parasitized by the Human Flea, *Pulex irritans* (7), though Dog Fleas, Cat Fleas and Rat Fleas will also transfer to man, and Human Fleas are often found on goats, pigs, badgers, foxes, and others. Human Fleas, which are found all over the world, can make jumps of a foot or more — a vast distance for so tiny an insect. They are less common than they used to be, because standards of cleanliness are higher. But wherever dirt is allowed to accumulate near human habitations, they breed in plenty, their fly-like, bristly larvae feeding in the dirt. They pupate in cocoons, and the adults on emerging attach themselves to the nearest host.

8-9 Thrips (Thysanoptera). The 180 or so British species are mostly tiny, inconspicuous insects, about many of which little is known. Most are active between spring and autumn, though some may be found on their host plants even in the depths of winter. The females of some insert smooth, soft-shelled eggs in the tissues of plants, whereas those of others attach rough, hard-shelled eggs to the surface of a plant. The almost transparent larvae hatch within 2 to 20 days, according to the species, and start sucking the plant sap; after a few days they change their skins and become coloured and tougher. Some species at this stage become carnivorous, attacking other Thrips and small insects. Some species mature quickly; others take several months, overwintering as larvae. They pupate in the soil or leaf-litter, during which stage they cannot feed, but can move slowly. The adults of many species, especially the predatory ones, move rapidly and fly actively; but some species are apparently incapable of flight, and others fly very little. But the flight of even the active ones is generally weak, and they are blown by the wind, often for great distances and in large numbers. Some species also jump. Though the adults of some attack other small insects, most suck the sap of plants, causing a silvery sheen where they have been.

The Onion Thrips (*Thrips tabaci*) (8) and the Grain Thrips (*Limothrips cerealium*) (9) are serious plant pests. Both are common in many parts of the British Isles, the females occurring throughout the year and hibernating in the winter. The Onion Thrips attacks plants such as ragwort and yarrow, as well as onions and related crops. There may be two or more generations in a year. The Grain (or Corn) Thrips attacks cereal crops and grasses from May to November. They often migrate in great numbers in the summer.

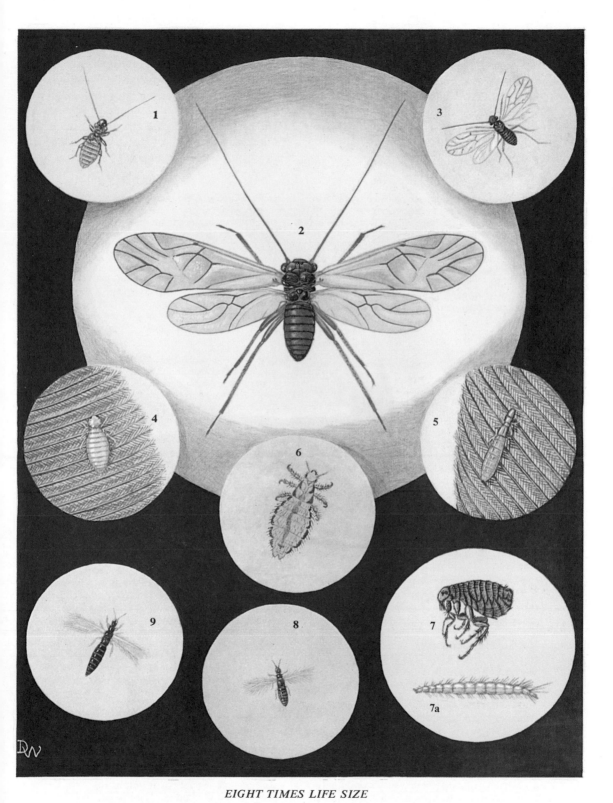

EIGHT TIMES LIFE SIZE

1 Trogium pulsatorium	2 Metylophorus nebulosus	3 Lachesilla pedicularia
4 Menopon gallinae	6 Pediculus humanus	5 Columbicola columbae
9 Limothrips cerealium	8 Thrips tabaci	7 Pulex irritans + larva

The true bugs (Hemiptera) have piercing and sucking mouthparts, forming a 'beak'. They have an incomplete metamorphosis — egg, nymph, and adult (*see* p. 198). Many have stink glands which they use for defence. There are about 1,630 species in Britain, divided into two groups, the Heteroptera (pp. 20–31) and the Homoptera (pp. 32–35). If the Heteroptera have wings at all, they are horny at the base but soft and flexible at the tips. The forewings fold over the more transparent hindwings when the insect is at rest. Shieldbugs are flat, broad insects with a conspicuous shield behind the head.

1 **Hawthorn Shieldbug** (*Acanthosoma haemorrhoidale*). These common bugs feed in large numbers on hawthorn berries. When these are not available, they eat the leaves or move on to other trees such as oak or whitebeam. The females lay batches of about 20 eggs on the undersides of leaves, and when these hatch in about 9 days, the young remain clustered around the shells until after their first moult. Then they begin feeding. There is one generation in a year and, like most Shieldbugs, they overwinter as adults. This species is not found in Scotland.

2 **Birch Shieldbug** (*Elasmostethus interstinctus*). A bug found wherever there are birches, especially with catkins, and also sometimes hazel and aspen. The female lays batches of some 24 eggs in May and June, and the new generation appears in August.

3 **Parent Bug** (*Elasmucha grisea*). This is one of the rare examples of an insect caring for its young. Early in June the female, having found a suitable leaf in a birch tree, rests awhile, and then lays on the underside of the leaf a diamond-shaped mass of 30 – 40 eggs, the right size for her body to cover conveniently. She remains with them till they hatch after 2 – 3 weeks. The young remain round the egg shells till after their first moult; then the Parent Bug leads the way to the birch catkins – the young following. While she stays with them, for how long it is not known, she protects them, putting herself between them and any danger. They live through the winter, though many are eaten by birds, particularly great tits.

4 **Pied Shieldbug** (*Sehirus bicolor*). A common insect in southern woodlands, waste ground, and hedgerows, but not found in Scotland or Ireland. The female lays 40 – 50 eggs in a shallow hole dug in the ground and, like the Parent Bug, watches over them, occasionally turning them with her 'beak'. When they hatch, it is thought that she leads them to the foodplant, white dead nettle or black horehound, and if she wanders off, they search for her. She apparently ceases to care for them once they start feeding. There may be two generations in favourable years, and the bugs hibernate, sometimes in the soil but usually under moss.

5 **Bishop's Mitre** (*Aelia acuminata*). A species found in well-drained grassland in southern England, but absent from Scotland and Ireland. The eggs are laid in two rows of about 6 on grass blades. After the second moult, the nymphs feed mainly on grass seeds, and on the continent can become a minor pest of cereals. They are eaten by ants and some kinds of beetles.

6 **Green Shieldbug** (*Palomena prasina*). This bug, common except in the north, is found, sometimes in great numbers, on hazel trees. The eggs are laid in midsummer, and the resulting adults hibernate, becoming darker, often reddish bronze, as the winter advances. In the spring they become bright green again.

7 **Sloe Bug** (*Dolycoris baccarum*). A widely distributed bug found on sloe and damson trees and many other plants, both young and adults feeding on the flowers and occasionally the fruits. Sometimes they feed on wheat, reducing the quality of the flour.

8 **Forest Bug** (*Pentatoma rufipes*). These are found mainly on oak and also on alder trees. They overwinter as nymphs in cracks and crannies high up in the trees, and tits eat a great number. In the spring they eat caterpillars as well as leaves, and are also sometimes very common in cherry trees at fruiting time, making the fruit uneatable. They are adult by July, and lay their eggs in August.

9 **Picromerus bidens.** These Shieldbugs, which have no common name, are useful to man as they eat caterpillars, even the distasteful larvae of the Cinnabar moth (p. 94), and leaf-eating beetles (p. 187). They are widespread in damp, lush vegetation, and overwinter as eggs. The female lays batches of about 30 eggs on the stems or leaves of shrubby plants in the autumn, and these hatch in early summer. The young at first feed on plants near the ground, and it is thought that they disperse and become carnivorous after the first moult. Some adults may also overwinter.

10 **Blue Bug** (*Zicrona caerulea*). A metallic blue insect, fairly common except in Scotland, and most often found on chalk, limestone, or sandy soils. The eggs are laid in small batches on a wide variety of plants in May to June. When they hatch, the young at first rest in groups with antennae touching. Later it is thought that they become at least partly carnivorous, but their life cycle is not fully known. The new generation appears from mid-July.

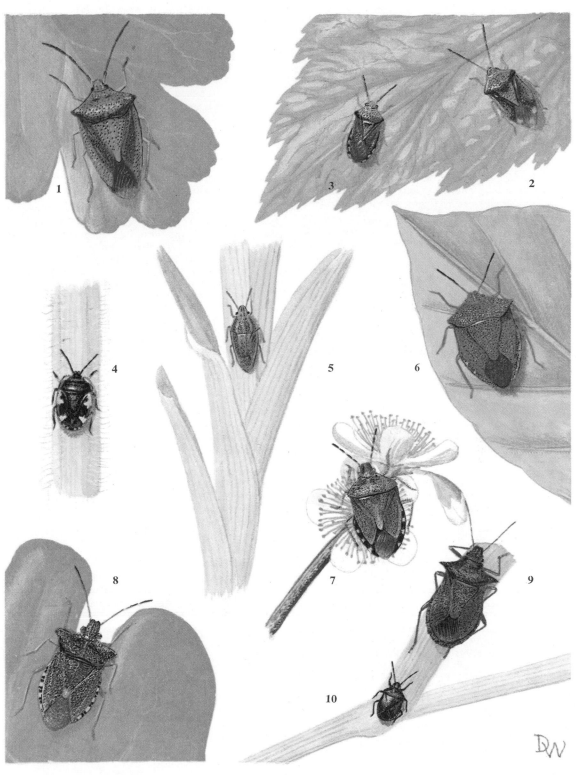

TWICE LIFE SIZE

1 ACANTHOSOMA HAEMORRHOIDALE   2 ELASMOSTETHUS INTERSTINCTUS   3 ELASMUCHA GRISEA
4 SEHIRUS BICOLOR   5 AELIA ACUMINATA   6 PALOMENA PRASINA
8 PENTATOMA RUFIPES   7 DOLYCORIS BACCARUM   9 PICROMERUS BIDENS   10 ZICRONA CAERULEA

There is a general note on the Hemiptera (Bugs) on p. 20.

1 Flatbugs (family Aradidae). Of the five species in Britain *Aradus depressus* is the most widespread, though absent in Scotland and scarce in Ireland. Its flattened body is adapted for living under the bark of trees, as well as on tree stumps and among woodland litter, where it feeds on fungi. Both young and adults are gregarious. Its life history is not fully known, but adults have been found in every month of the year. Possibly a bug living under bark is less influenced by weather conditions than those living more in the open.

2-5 Squashbugs (family Coreidae). A well-known American species, *Anasa tristis*, attacks squashes and other gourd plants — hence the name. The British species are dull brownish, although the upper part of the abdomen, normally covered by the wings, is often bright coloured. The antennae, which are very large in the nymph stage, are 4-segmented, and are sometimes used as levers, should the bug fall on its back.

Syromastus rhombeus (2) is found in dry, sandy places and chalk pits in southern England and Ireland. feeding especially on sandworts and spurreys. Little is known of its life cycle except that adults appear in May, and that they overwinter in grass-tufts and at the base of trees.

Corizus hyoscyami (3) is found almost exclusively in sandy areas on the south and west coasts and parts of the south-east Irish coast. The nymphs, which especially frequent restharrow but also many other plants, are quite unlike the adults, being very hairy and yellowish-brown with faint pink spots. The new generation appears in September and overwinters in moss or on juniper or pine trees.

Rhopalus subrufus (4) is mainly found south of the Thames, but also sometimes in Norfolk, in clearings in woodlands, especially where St. John's wort grows. The eggs are laid in May, the adults appearing in late summer and overwintering.

Myrmus miriformis (5). This species is common in open, damp grassland and sometimes stagnant grass heaths, and it feeds on various grasses. The female lays her eggs in ones or twos from mid-July, gluing them on to grass blades or heather, and the eggs overwinter. They hatch from May onwards. There are two colour forms, a green and a brown, though the female is always green. The females usually have incomplete wings and cannot fly. The males may have either complete or incomplete wings, the reason for this variation being unknown.

6 Nettle Groundbug (*Heterogaster urticae*). Groundbugs of the family Lygaeidae (Nos. 6–9) are mainly seed-feeders and are small, black or dark-brown insects. This species is common in lowlands in England south of Yorkshire. It spends its life on stinging nettles in warm situations on alkaline wasteland. In June and July the adults can sometimes be seen congregating on the upper parts of the plants. The females lay their eggs in the ground at the base or occasionally on the lower leaves; they cover them with a secretion which hardens. The new adults appear in September and overwinter beneath bark or in hollow, woody stems.

7 Reedmace Bug (*Chilacis typhae*). A southern English species, rare in the north and uplands. Adults are found in the heads or litter of bulrushes (reedmace), where they spend the winter. The eggs are laid on the seeds or pappus. The red-brown nymphs mature from mid-July onwards, sometimes there is a partial second generation, the nymphs of which overwinter with the adults.

8 European Chinchbug (*Ischnodemus sabuleti*). Unlike most insects, this species has increased in numbers in the last 40 years in the south-eastern counties of England, and is now often to be found in thousands congregated on the heads of various grasses and reeds. Both adults and nymphs hibernate in leaf sheaths and clumps of grasses and reeds and start feeding again about mid-April. The adults begin to lay eggs in groups of 3 or 4 in grasses in late May, but the nymphs do not mature and start laying until into July. The progeny of these do not reach maturity before winter, and so hibernate as nymphs. In the warm weather the adults often then fly from the reedy swamps to dry sunny fields and feed on oat grasses and others. About half the Chinchbugs of a season have fully developed wings, and the rest have incomplete wings, most of them with very much reduced forewings.

9 Trapezonotus arenarius. A common species with no English name, most usually found on sand-dunes, though little is known about its foodplants. There are two other similar but rather larger varieties, which are now considered separate species.

10 Pine-cone bug (*Gastrodes grossipes*). A bug of Scots pine trees and also Norway spruce and other pines. The adults spend the winter inside old cones, becoming active whenever the temperature rises. They emerge in the spring, and the females lay their eggs in rows along the pine needles. These hatch in late April or later in a cold year. In the north they may hatch up to a month later. They take about 8 weeks to become adult, and by early August most have matured.

THREE TIMES LIFE SIZE

1 ARADUS DEPRESSUS   2 SYROMASTES RHOMBEUS   3 CORIZUS HYOSCYAMI
4 RHOPALUS SUBRUFUS   5 MYRMUS MIRIFORMIS   6 HETEROGASTER URTICAE
7 CHILACIS TYPHAE   8 ISCHNODEMUS SABULETI   9 TRAPEZONOTUS ARENARIUS
10 GASTRODES GROSSIPES

23

There is a general note on the Hemiptera (Bugs) on p. 20.

1-2 Lacebugs. These small, reticulated insects of the family Tingidae all have lace-like wings, and are plant feeders. The Rhododendron Bug (*Stephanitis rhododendri*) (1), first found in Britain in 1901 in Surrey, is now found over much of England, southern Wales, and eastern Scotland. Adults appear from the end of June to October. They pair in July, and the females lay their eggs in rows or long clusters embedded in the undersides of rhododendron leaves near the midrib, and cover them with a secretion which dries to form a brown scab. The young at first are gregarious, but become more solitary as they grow older. This bug can be detected on rhododendrons by whitish mottlings on the upper leaf surfaces or by brown spots.

The Creeping Thistle Lacebug (*Tingis ampliata*) (2) is common in southern England and the midlands. In May the adults leave their overwintering quarters amongst mosses and litter and move on to creeping thistles, usually rather less than ten to each plant. From late May until early July, the females lay white eggs in the plant stems, so that only the tops of the eggs show. The young hatch in late June and July, and are not very active at first, but as they grow older they move about the plant freely. The new generation appears in August and early September.

3 Assassin Bugs (family Reduviidae). These, as their name indicates, are carnivorous, mostly blood-suckers. The Heath Assassin Bug (*Coranus subapterus*) is common in Britain, but not Ireland, on open heaths and sand-dunes. Normally it overwinters in the egg stage. Adults are found from July to October, and they mate in autumn. The dark-brown eggs are laid in cracks and crevices in moss and litter, and hatch the following April or May. This species hunts on the ground for insects and spiders, and catches its prey with its 'jack-knife' front legs and stout, curved 'beak'. The young take about two months to become adult. Both young and adults stridulate loudly when touched.

4-6 Damsel Bugs (family Nabidae). These are closely related to Assassin Bugs, and most are very small. Field Damsel Bugs (*Nabis ferus*) (4) are found on dryish ground amongst grasses everywhere in the British Isles. The adults overwinter, and the eggs, laid in grass stems in late May and early June, hatch in about 10 days. Both adults and young are predacious, probably detecting their prey by scent and movement, and then using the front legs to seize them and the 'beak' to spear them. Their prey includes meadow plant bugs, leaf-hoppers, aphids, and various caterpillars. The fully-winged adults appear in August and September and are active fliers, migrating to fresh habitats, particularly after hibernation.

The Ant Damsel Bug (*Himacerus mirmicoides*) (5) occurs in dry sunny situations in southern England, Wales, and Ireland, preferring sloping grassy places. It is rare in the north and absent from Scotland. The adults hunt mainly at night, even as late as November, preying on aphids, butterfly eggs, and other bugs. They over-winter in hedgebottoms. The eggs, laid in stems of grasses in late spring, hatch between mid-June and early August. The new generation begins to appear in August. The young strongly resemble small black ants and have been found in ants' nests.

The Marsh Damsel Bug (*Dolichonabis limbatus*) (6) is common in damp meadows and rank vegetation throughout the British Isles. The eggs are laid from mid-August until early November in grass stems where they overwinter and hatch in late May. The young at first spend most of their time near the tops of grasses; but later, like the adults, they stay closer to the ground. The new generation appears from early July onwards.

7 Common Flower Bug (*Anthocoris nemorum*). This is one of the most common bugs in the British Isles. The adults appear in southern England in late June and July; the second generation appears in September. They overwinter under bark or in litter, and in early spring colonize a wide range of plants. They prey on small insects and are occasionally herbivorous. The white eggs are inserted in the undersides of leaves, and hatch in late May and June. This insect is useful to man as it feeds on red-spider mites, greenfly, and other pests on many deciduous trees.

8 Bedbug (*Cimex lectularius*). A closely related species to the Common Flower Bug (7) which sucks the blood of humans, poultry, mice, and many zoo animals, and in a suitable temperature breeds continuously. Bedbugs were first recorded in England in 1503; in 1939 it was calculated that four million people in Greater London alone were troubled by them, and they have been recorded from most British towns. Now they are rarer, because they can be efficiently controlled by DDT. During the day, they hide in crevices or behind wall-paper and pictures, or in bedding, and at night emerge to seek a meal of blood. They locate their hosts by scent and temperature, but only when within an inch or so of them. Bedbugs have not been known to transmit disease in Britain, although they sometimes carry disease-causing organisms.

9 Bracken Bug (*Monalocoris filicis*). A common and widely distributed plant bug (*see* p. 26) throughout the British Isles on bracken and ferns, where it feeds on the spore cases. The adults overwinter, emerging in late April and May. The female lays a mass of green spore-like eggs from late May to the end of June. The adults of the new generation, which appear in late July and August, are fully winged and fly readily until they hibernate in the autumn.

10 Dicyphus epilobii. This plant bug is common on great hairy willow herb throughout the British Isles from mid-July till late autumn. It overwinters in the egg stage. The young are pale green, with the basal part of the 'feelers' red. Those occasionally found as late as mid-August may be a second generation.

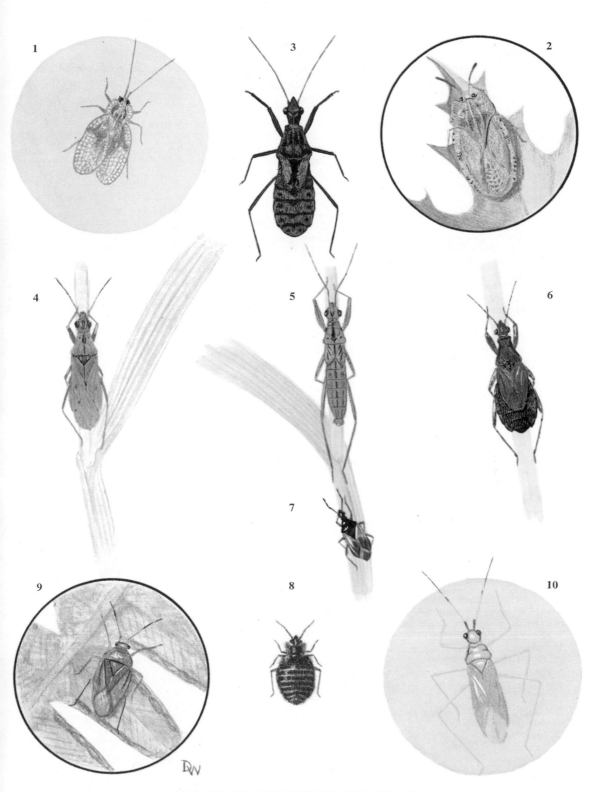

FOUR TIMES LIFE SIZE (IN CIRCLES × 7)

1 STEPHANITIS RHODODENDRI   3 CORANUS SUBAPTERUS   2 TINGIS AMPLIATA

4 NABIS FERUS   5 HIMACERUS MIRMICOIDES   6 DOLICHONABIS LIMBATUS

7 ANTHOCORIS NEMORUM

9 MONALOCORIS FILICIS   8 CIMEX LECTULARIUS   10 DICYPHUS EPILOBII

25

PLANT BUGS (*Miridae*)

The members of this large family, often called capsids, live on plants but vary widely in shape, structure, and diet. Most feed on plants, but a few are carnivorous. (*See* also pp. 24 (9, 10) and 28.)

1 Pilophorus cinnamopterus. This species occurs south of a line from the Wash to Glamorgan, and lives on Scots pine, feeding on the leaves and resinous sap, as well as on black aphids. It overwinters as an egg. The reddish young mature by mid or late July.

2 Delicate Apple-capsid (*Malacocoris chlorizans*). A common bug throughout the British Isles on hazel, and also on elm, apple, and other trees. It lays eggs in small batches on the undersides of leaves near the main vein. Those laid in autumn hatch the following May into pale-green young, and the adults appear from mid-June onwards. A second batch hatches in July and August, and the second generation adults appear from late August onwards. Both young and adults feed on mites and mites' eggs, and sometimes on the host plant.

3 Cyllecoris histrionicus. A widespread species throughout the British Isles on oak. The eggs are laid in summer, often 5 or 6 closely packed together in a row in bark cracks of young wood, where they overwinter. The young, pale greenish-white with dusky markings, hatch the following May, and the first adults appear in June. Both young and adults feed on unopened catkins, very young acorns, mites, aphids, etc., the adults becoming more predacious as they grow older.

4 Heterotoma merioptera. This bug is abundant in southern England on rank vegetation such as nettles, and on various shrubs and trees; but it is rare in the north. The young are red, hatching from overwintered eggs in late May and early June. The adults appear from late July onwards, and feed on aphids and other small animals, and on buds and various unripe fruits. This bug can be identified by the greatly-enlarged second joint of the antennae.

5 Mecomma ambulans. A common bug in rank vegetation at the margins of woods and in clearings all over the British Isles. It lays yellow eggs in rushes, where they overwinter and hatch into greenish-yellow young in the following late May. They are probably predacious, but also feed on the juices of rushes and sedges. The new generation appears from July to September. The fully-winged males look much like ichneumon flies, the females are usually short-winged.

6 Pithanus maerkeli. This bug occurs throughout the British Isles in damp, grassy meadows with rushes. The eggs are laid in July and August in rush and grass stems and hatch the following spring. The dark-brown to black young, though reared on plants, later certainly eat leaf-hoppers (p. 33) and seem to be omnivorous.

7 European Tarnished Plant Bug (*Lygus rugulipennis*). Found throughout the British Isles on a wide variety of herbaceous plants, including nettle, dock, and clover, this is sometimes a pest of lucerne, potatoes, chrysanthemums, and raspberries. It attacks the young shoots, flowers, and fruits, and produces extensive white spots on the leaves. The adults overwinter in dead leaves and litter; many die, and the rest reappear in March and April, and lay eggs in May in the unopened flowerbuds and stems of the host plant. The young, green with two rows of black spots, become adult in late July. By September there is a second generation — but probably not in Scotland. These bugs are most common in autumn, when they fly about a great deal.

8 Liocoris tripustulatus. A bug found on nettles throughout the British Isles. The adults overwinter, and by June most of the males have died, while the females are laying their eggs in the leaf stalks just below the leaf. These females often survive until the adults of the new generation appear. The young are green with red-brown markings. All stages are entirely plant-feeding, attacking buds, stems, flowers, and fruits. Before hibernation the adults are pale yellowish brown, but by spring are deep chocolate with orange spots.

9 Common Green Capsid (*Lygocoris pabulinus*). This bug, a minor pest of recent years, is abundant throughout the British Isles and attacks a wide variety of plants. The eggs are laid in September in young twigs of hawthorn, apple, currant, cherry, or lime. The green young hatch the following spring and, after moulting twice, move to herbaceous plants — particularly nettles, creeping thistles, groundsel, dandelions, black nightshade, potatoes, and strawberries. They mature in June. These adults lay eggs in late June and July in the stems and leaf-stalks of their foodplants. The second generation is mature by September, and after pairing returns to the twigs of the hawthorn and winter hosts. When this bug occurs on raspberry, as it does in Scotland, or rose, or elder, it often remains on it throughout the year. Of recent years the young of the first generation have caused damage to the leaves and fruit of currants and gooseberries, and also to pears, the damaged fruits of which develop irregular distorted shapes and form corky tissue. The second generation may severely check the plant growth of blackberries.

10 Apple Capsid (*Plesiocoris rugicollis*). This bug used to feed on various willows and on bog myrtle; but in 1914 it was noticed in large numbers on apple trees, where it became a severe pest. It can be controlled by DDT, and is now sparsely distributed throughout the British Isles. The eggs, laid in the bark of young shoots during late June and July, overwinter there, and the yellowish-green young hatch the following April and early May. At first they feed on the unfolding leaves but, as soon as the petals fall, they attack the young fruits, causing little brown spots which extend into rough, corky patches as the apples grow. The adults appear in early June to the end of July. There may sometimes be a second generation.

FOUR TIMES LIFE SIZE

1 PILOPHORUS CINNAMOPTERUS   3 CYLLECORIS HISTRIONICUS   2 MALACOCORIS CHLORIZANS

4 HETEROTOMA MERIOPTERA   6 PITHANUS MAERKELI   5 MECOMMA AMBULANS

7 LYGUS RUGULIPENNIS   8 LIOCORIS TRIPUSTULATUS

9 LYGOCORIS PABULINUS   10 PLESIOCORIS RUGICOLLIS

27

Other plant or capsid bugs (Miridae) are described on p. 26.

1 Miris striatus. A very large plant bug found most frequently on oak, but also on hawthorn or apple. Although widely distributed, it is never found in large numbers. The eggs are laid in the bark of young wood in June and hatch the following year in April and May. Both the young and the adults are unusual for this family in being mainly predacious, feeding on aphids, caterpillars, moth pupae, and bug eggs.

2 Potato Capsid (*Calocoris norvegicus*). A common bug over most of the British Isles, living on mixed vegetation in damp, sheltered places, and feeding on the buds, flowers, and unripe fruits, particularly of nettles, thistles, and clover. Occasionally, it is a minor pest of potatoes, carrots, and chrysanthemums. The eggs are laid in July and August in cracks in stems, where they overwinter. The green young hatch the following May, and by the end of June, the adults appear.

3 Lucerne Plantbug (*Adelphocoris lineolatus*). This bug is widely distributed throughout the British Isles. It feeds on the young leaves, stems, flowers, and unripe fruits of vetches, clover, restharrow, and other legumes. The adults also feed on the flowers of scentless mayweed and mugwort, and in Britain they are occasionally pests on onions, beet, and chrysanthemums. In America, Poland, and southern Russia they are major pests of lucerne and clovers. The eggs are laid from mid-July onwards, either singly or in batches of nearly fifty, in the stems of the food plants, where they overwinter, hatching the following May. The dark-green young become adult by early July.

4 Phytocoris tiliae. This species is common throughout the British Isles on many deciduous trees, particularly oak, ash, lime, and apple. The eggs are laid in young wood in mid-summer, and the green-and-black young hatch the next year in early June. Adults appear from late June onwards. This, like *Miris striatus* (1), is almost entirely predacious, feeding on caterpillars, red-spider mites, and other small animals.

5 Capsus ater. There are two colour forms of this bug: one entirely black, and the other partly reddish brown. It is found throughout the British Isles in long grasses, particularly perennial rye and couch grass. The eggs are laid between the sheathing leaf-base and the stem, in batches of up to thirty, in late June and July. The purple-red young hatch the following May and June and feed on the lower parts of the stem, not on the fruits and buds as many plant bugs do.

6 Capsodes gothicus. Though not very common, this species appears throughout England and has been found in Glamorgan. It occurs in thick vegetation, particularly on large birdsfoot trefoil in marshy places.

The eggs, laid in July, hatch the following late spring. The young are black and deep red with long, black hairs. The adults begin to appear in the last fortnight of June.

7 Stenodema calcaratum. These plant bugs, common throughout the British Isles on grasses, overwinter as adults in litter and other sheltered places and appear again in April. They lay eggs in June and July, usually singly, in spikelets of grasses, and the young mature from the middle of August.

8 Meadow Plantbug (*Leptopterna dolabrata*). These bugs abound throughout the British Isles in fairly moist, grassy places, particularly on timothy and couch grass, meadow foxtail, and Yorkshire fog. The eggs are laid in summer in the lower parts of grass stems, and partially develop during late summer and autumn; they become inactive during the winter, but with the warmer spring weather, they complete development, and in May hatch into greenish young with black markings. These feed on most parts of the plant. The adults appear at the end of June, the males maturing first, and are plentiful in July, but very rare by September.

9 Common Shorebug (*Saldula saltatoria*). This is probably the most common and widespread British shorebug, and lives around the margins of ponds, ditches, and semi-stagnant streams and lakes, on firm mud or silt. The adults overwinter under tufts of grass or any suitable shelter, becoming active on warm days when they run over the mud hunting their prey. The adults pair between late April and June, and the eggs are laid in cracks in mud or in low vegetation. These have produced new adults by July, and in some years in the south there may be two generations.

10 Marine Bug (*Aepophilus bonnairei*). A very local bug found in a few sea-side localities along the south coast from the Isle of Wight westwards, N. Devon, two places in Wales, the Isle of Man, Dublin, Mayo, Waterford, Cork, and Galway. This 'Atlantic' distribution suggests that there may once have been a continuous shoreline between Brittany and the west coast of Ireland. The Marine Bug usually lives in the cracks of slate and shale, or in holes in stones between high and low water marks, which are often flooded, and have entrances curtained with seaweed. To find this tiny bug people lever open with crow bars the cracks and fissures on the least exposed side of the rocks, when the tide is out. The bugs are predacious, but it is not certain what their prey is. They hunt only when the rocks are above water, and otherwise lead a rather inactive life in family groups. Adults and young are found all the year round, so breeding is probably more or less continuous in the cracks and fissures where the temperature and humidity remain fairly constant throughout the year.

FOUR TIMES LIFE SIZE

1 MIRIS STRIATUS   2 CALOCORIS NORVEGICUS   3 ADELPHOCORIS LINEOLATUS

4 PHYTOCORIS TILIAE   5 CAPSUS ATER

6 CAPSODES GOTHICUS   7 STENODEMA CALCARATUM

9 SALDULA SALTATORIA   10 AEPOPHILUS BONNAIREI   8 LEPTOPTERNA DOLABRATA

29

Water bugs are a group of several families adapted for living either in the water or on its surface All, except *Corixa* (No. 10), are predacious.

1   **Pondweed Bug** (*Mesovelia furcata*). A bug found on the surface film of small ponds with floating vegetation, mainly south of a line Suffolk to Somerset. It lays its eggs in September in the stems of floating plants, which sink to the bottom of the pond in winter. When the young hatch in spring, they swim up and break through the surface film, finding their prey among surface-dwelling animals. The new generation of adults appears in July and August.

2   **Water-measurer** (*Hydrometra stagnorum*). A fairly common bug, which walks slowly amongst vegetation on the surface film of stagnant or sluggish water. The adults overwinter, pair in the spring, and the female lays long, narrow, and seed-like eggs, stuck sideways on plant stems at water level. Adults from these appear in June and July. The Water-measurer detects its prey by vibrations and spears it through the surface film.

3   **Water-cricket** (*Velia caprai*). Despite its name, this widely distributed bug is silent. It is usually found in streams, or in ponds with some slight flow, where it locates its prey both by sight and by reacting to ripples. The eggs, laid in late spring on moss, hatch and mature in about two months, the adults appearing in August.

4   **Common Pondskater** (*Gerris lacustris*). This is probably the most widely distributed British bug. Pondskaters use the front legs to grasp their prey, the middle legs as oars to row themselves over the water surface, and the hind legs as rudders. They may use their very large eyes as well as their vibration detectors to locate their prey. The adults seek sheltered places away from water during the winter, and return to ponds, lakes, or streams in spring, and lay their eggs in May. These have become adult by July, and by mid-August there is a second generation of adults.

5-6 **Water Scorpions** (family Nepidae). The two British species, though unalike, both have long 'tails'. These are not stings, as often thought, but breathing tubes with which they penetrate the surface film and bring air to the insects lying submerged in wait for their prey. The painful sting is made with the beak. Water scorpions do not hunt actively, but wait stealthily for insects, tadpoles, and small fish to come within reach of their 'jack-knife' front legs. They harmonize well with their surroundings and protect themselves by concealment.
Nepa cinerea (5), is the more common of the two, being found in weedy stagnant ponds and shallow lakes almost everywhere except in northern Scotland. In late spring, the females insert their eggs into the stems and leaves of plants just below the water surface. The young take about two months to mature. The adults are very poor swimmers, preferring to walk in the water, and are usually flightless.
Ranatra linearis (6), usually called the Water Stick-

insect, is found south of a line Norfolk to S. Wales, usually in water no more than 3 feet deep, in ponds with woody vegetation and rubbish. The adults are active throughout the winter, except when it is very cold. In spring the females insert eggs in plant stems, usually pondweed or water plantain, and these are supplied with air by long filaments which come to the surface. They take longer to hatch in a cold spring. Water Stick-insects can swim quite well and, if conditions in the pond become unsuitable, they crawl out, open their wings with the aid of their hind legs, and fly elsewhere.

7   **Saucer Bug** (*Ilyocoris cimicoides*). A very common bug in muddy ponds, dykes, and occasionally in streams south of a line Nottingham to Staffordshire. It swims actively, searching the bottom of the pond for its prey — water lice and other aquatic animals. Although it has wings, it cannot fly but walks from one pond to another. In the spring the female lays eggs in the stems of water plants, and these produce adults by August, which seem to hibernate in mud at the bottom of ponds during the winter.

8   **Water Boatman or Back Swimmer** (*Notonecta glauca*). Found over most of the British Isles, these lively bugs swim on their boat-shaped backs and can fly, and dive. They hunt tadpoles and diving beetle larvae, which they detect by sight and vibrations in the water. They destroy a lot of fry in trout hatcheries. They can inflict a painful sting with their beaks. The adults pair from mid-winter to May, and eggs are laid from February onwards in the stems of water plants. The nymphs take about two months to mature.

9   **Plea atomaria.** A tiny bug common in weedy ponds, lakes, and slow rivers, especially where there is Canadian pondweed, throughout England and Ireland. It is rare on high ground in the north, and absent from most of Scotland. The females insert eggs in water plants in July and August, and the new adults begin to appear in September and overwinter.

10 **Corixa punctata.** Widespread over the British Isles, this bug, sometimes called a Water Boatman, lives in ponds. The flat, hairy ends to the forelegs act as sieves, and the bug eats everything, plant or animal, which is caught in the sieve. It swims and flies well, but is much less vigorous than the proper Water Boatman. From late January to March the females lay eggs, which they glue to water plants, and which hatch according to the temperature — taking on average about three weeks. The young become adult in three to four months. Male Corixids, when courting, stridulate or 'sing' by rubbing hair patches on the inside of their front legs against the side of their beaks. Certain tiny bivalve animals sometimes attach themselves to the legs of the Corixid and may be carried by it to new ponds, which they then colonize.

TWICE LIFE SIZE (IN CIRCLES × 6)

2 HYDROMETRA STAGNORUM

1 MESOVELIA FURCATA	3 VELIA CAPRAI	4 GERRIS LACUSTRIS
6 RANATRA LINEARIS	7 ILYOCORIS CIMICOIDES	5 NEPA CINEREA
10 CORIXA PUNCTATA	8 NOTONECTA GLAUCA	9 PLEA ATOMARIA

These insects, as well as those described on p. 34, belong to a sub-order of Hemiptera (*see* p. 20) called Homoptera or 'like-winged', so called because both pairs of wings are alike in being almost transparent. At rest, they usually roof their backs. The Homoptera are very varied, but mostly small, all vegetarian, and all land-dwellers. The cicadas and hoppers described on this page are all extremely active, leaping bugs.

1 Cicadas (family Cicadidae). This large family, well known in hotter lands for their 'song', has only one, extremely rare and local, British representative, the New Forest Cicada (*Cicadetta montana*). The male Cicada 'sings' to attract the female. In his abdomen, just beyond the thorax, is a pair of drum-like organs, to the centre of which is attached a powerful muscle. Each time this muscle contracts or relaxes, a sharp click results. The rapid repetition of clicks reverberating through an 'air-chamber' sounds like a continuous song.

The female lays her eggs in dry twigs on trees, and when they hatch, the young fall to the ground and burrow down to suck sap from the roots, staying there usually for three years. Then, they come to the surface, climb the undergrowth, the skin splits down the back, and the adult emerges in June or July.

2 Tree-hoppers (family Membracidae). This large and mainly tropical family, with many curious forms, has only two British representatives, one being the Horned Tree-hopper (*Centrotus cornutus*). The adults have large heads. They may be beaten from low branches in oak woods, and occasionally from lower shrubs, from April to August in many parts of Britain. The young live on oak foliage and in the dead leaves below.

3-5 Frog-hoppers (family Cercopidae). These little bugs are frog-like in appearance and in jumping ability. The young of most species secrete frothy masses around themselves. The Common Frog-hopper (*Philaenus spumarius*) (3) is the well-known 'cuckoo-spit insect' of British gardens and hedgerows. The females lay eggs in crevices in dead stems in late October and November, and these hatch in early spring. The young (nymph) ascends the young stem of a plant, finds a suitable site (generally in the axil of a leaf), and begins to suck the sap. It protects itself from enemies, and also perhaps from the sun, with a mass of white froth, formed by a secretion from the abdomen blown into bubbles by air from a nearby cavity. Masses of froth, from which the bug has often moved on, may be found in spring, especially on hawthorn and sorrel. The adults appear from June to November.

The Alder Frog-hopper (*Aphrophora alni*) (4) is found mainly on alders and sallows, but also on many other shrubs and trees, over most of the British Isles. The adults, which have two distinct white patches on the margins of the forewings, appear from May until October.

Red and Black Frog-hoppers (*Cercopsis vulnerata*) (5) the largest of British Frog-hoppers, are found locally in England, especially in the south, from April to August in or near woods, and sometimes on willows in marshy ground. They are rare in Scotland. Their conspicuous warning coloration warns predators that they are distasteful. The young pass the winter underground (sometimes a foot deep), and in spring they attach themselves to the roots of grasses, bracken, or docks, secreting a mass of froth around themselves as they suck the sap. When mature, they leave the froth and move up on to the plants.

6-9 Leaf-hoppers (family Cicadellidae). A very large, varied family with 253 British species, some of which are pests. Both the adults and young of the Eared Leaf-hopper (*Ledra aurita*) (6) have a flattened appearance, and their mottled colour-scheme blends in well with the lichen-covered branches of the oak trees on which they live. Their name comes from a pair of ear-like projections on the front edges. The adults appear from May to September in the southern half of England. The Green Leaf-hopper (*Cicadella viridis*) (7) is commonly found from July to October on grasses and rushes in marshy places throughout the British Isles. Also common throughout the British Isles is *Evacanthus interruptus* (8), formerly a pest of English hop fields. The adults may be found from June to October on various grasses and bushes.

The Rhododendron Leaf-hopper (*Graphocephala coccinea*) (9) was introduced into Britain from America about 1935, and was first found at Chobham, Surrey. Since then it has spread as far afield as Dorset, Cheshire, and Monmouth. Both young and adults feed on the leaves of wild and cultivated rhododendrons and are responsible for spreading a fungus disease Bud Blast, which causes the rhododendron flower buds to turn brown and die off without opening. Adults of these Leaf-hoppers can be seen on rhododendrons from April to October.

LIFE SIZE
1 CICADETTA MONTANA   2 CENTROTUS CORNUTUS
3 PHILAENUS SPUMARIUS
4 APHROPHORA ALNI   6 LEDRA AURITA
5 CERCOPSIS VULNERATA
7 CICADELLA VIRIDIS   8 EVACANTHUS INTERRUPTUS   9 GRAPHOCEPHALA COCCINEA

There is a note on the sub-order Homoptera on p. 32. The species on this page are all pests which pierce and suck the juices of plants. Most can be controlled by spraying.

1-2 Jumping Plant-lice (family Psyllidae). These are small aphid-like bugs, but with long antennae and legs, whose young (nymphs) may cause galls on plants. The rather uncommon Rush Sucker (*Livia juncorum*) (1) causes conspicuous 'tassel galls' on rushes from July to October, each gall containing a single nymph. The Apple Sucker (*Psylla mali*) (2), once a serious pest of apple orchards, has winged adults which lay eggs in bark crevices on the fruiting spurs in autumn. In April the young hatch and feed on the sap, causing foliage and buds to shrivel. The first adults appear by mid-May.

3-6 Aphids (family Aphididae). This family, which includes the well-known Greenfly and Blackfly, suck the sap of plants, which they then excrete as a sweet, sticky fluid, loved by ants (*see* p. 150). Some species also secrete a waxy substance through tubes at the end of the abdomen. Many have very complicated life histories, of which the history of the Black Bean Aphid (*Aphis fabae*) (3) is typical. This Blackfly or 'Black Blight' damages crops and garden plants from May to July. It occurs in vast numbers in two forms, winged and wingless, and has two generations a year. In late summer, winged females from the summer host plants, beans, sugar beet, docks, etc., migrate to spindle trees or sterile guelder roses, where they produce wingless, egg-laying females. In early September, winged males fly over to these and pair with them. The females lay minute, black eggs and then die off. In spring wingless females hatch out which, without mating, produce living young (not eggs). These colonies in turn produce the winged forms, which return in early summer from the spindle to the beans and other summer host plants, where further winged and wingless generations are produced, and the cycle is started over again.

The Currant Blister Aphid (*Cryptomyzus ribes*) (4) lays eggs on the hardwood of currant bushes during autumn. The young hatch in spring and feed on the under-surfaces of the young leaves, causing them to blister and turn bright red. Winged forms are produced in July, which migrate temporarily to lettuce, knotweeds, or various dead-nettles.

The Woolly Aphid or American Blight (*Erisoma lanigerum*) (5) attacks trees, especially apples, sucking the sap and causing round swellings. It cannot effectively be controlled by sprays. It was first recognized in Britain in 1766. Wingless females overwinter in bark crevices and produce a succession of broods throughout summer and autumn, some of them winged. The conspicuous masses of a white, fluffy substance conceal whole colonies of young Aphids.

The Spruce Pineapple Gall Aphid (*Adelges abietis*) (6), has two distinct races. One has a one-year life cycle and remains throughout on the spruce tree. The other has a two-year life cycle: the winged adults migrate in August to larch trees, where they lay eggs which overwinter; winged adults hatch out in May and June, and return to the spruces, where the females lay eggs which produce wingless females; these overwinter and then lay eggs which start the cycle again. Both races cause pineapple-shaped galls to form at the bases of the spruce needles.

7-8 Whiteflies (family Aleyrodidae). These tiny, actively-flying bugs are covered with a white, powdery wax, and look like minute moths. The Cabbage Whitefly or Snowy Fly (*Aleyrodes brassicae*) (7) spoils cabbage and related plants. In autumn, when disturbed, the adults rise in clouds and settle again like snow. They breed throughout the summer. The young congregate on the undersides of leaves, their white, waxy, yellow-spotted scales discolouring the leaves, and their excretion spoiling them for food.

The Greenhouse Whitefly (*Trialeurodes vaporariorum*) (8) attacks cucumbers, tomatoes, and many greenhouse plants, making them lose colour and covering the leaves and fruit with a sticky honeydew. In summer, after mating, the females lay circular groups of eggs on the undersides of young leaves, and the pale-green, scale-like young feed there almost motionless; after about a month, they become adults. Whitefly can be controlled by introducing a species of Chalcid Wasp (p. 146), the grubs of which eat the Whitefly young. The Greenhouse Whitefly is found at all stages throughout the summer. Out of doors it does not survive British winters.

9-10 Scale-insects and Mealy-bugs (family Coccidae). These are very small insects, the delicate males possessing only one pair of wings, and the females none. Scale-insects are covered by a protective scale-like secretion. The females often lack, or almost lack, legs and antennae, and remain throughout life attached to their food-plant. Mealy-bugs are coated with a waxy secretion, and the females can move.

The Mussel Scale (*Lepidosaphes ulmi*), a pest of orchards, plasters the stems and branches of trees, especially apple, with brown, shell-like structures looking just like tiny mussels. Under each is a female, which lays her eggs there and then disintegrates. Tiny, wingless nymphs hatch out about late May, wander about for a few days, and then drive their 'beaks' into the bark and remain there feeding. Soon, long, white, waxy filaments form on their backs and, along with their own cast skins and excrement, harden into the scale-like covering. They soon lose their legs and become small, yellow, fleshy grubs, which mature in late summer. The males, which are rare, emerge as small, active, fly-like, two-winged insects which soon die after pairing. Only a small proportion of females mate; the rest, like aphids, reproduce asexually.

The San José Scale (*Quadraspidictus perniciosus*) (9) is a serious pest of fruit trees, and is an example of a pest which has reached Britain on imported crops. Unlike some, however, it cannot survive out of doors, but has been found in glasshouses.

The Oyster-shell Scale (*Q. ostreaeformis*) (10) sucks the sap of fruit trees, and occasionally birch, poplar, or horse chestnut. Its life cycle is in general similar to the Mussel Scale, but it overwinters as a nymph.

INSECTS IN CIRCLES × 6

1 LIVIA JUNCORUM   3 APHIS FABAE   2 PSYLLA MALI
4 CRYPTOMYZUS RIBES   5 ERISOMA LANIGERUM
6 ADELGES ABIETIS   8 TRIALEURODES VAPORARIORUM   7 ALEYRODES BRASSICAE
9 QUADRASPIDICTUS PERNICIOSUS   10 QUADRASPIDICTUS OSTREAEFORMIS

35

ALDER-FLIES, LACEWINGS, AND SCORPION-FLIES

There are three separate orders of insects shown here, with names meaning 'big wings' (Megaloptera, 1–2), 'nerve' or 'lace wings' (Neuroptera, 3–8), and 'long wings' (Mecoptera, 9–10). All the species described on this page are common or fairly common. All, except No. 9, are carnivorous; Lacewings especially attack aphids, the larvae draining the aphids dry with their special sucking jaws. All, except Nos. 9 and 10, have a rather weak and slow flight, and rest with their transparent wings closed over the abdomen.

1 Alder-fly (*Sialis lutaria*). This is the paler and the more common of the two British Alder-flies. It is usually found resting in large numbers on waterside vegetation or flying, rather weakly, on sunny days in late spring. The eggs are laid in large batches on plants or other objects at the water's edge, from which the brown larvae drop into the water and feed voraciously on other small creatures in the mud for up to two years. Then they leave the water and pupate in cells in the soil near by.

2 Snake-fly (*Raphidia notata*). All four British Snake-flies, of which this is the largest, have long, neck-like extensions of the thorax; the females have long, needle-like ovipositors. This species is found from May to July in oak and pine woods, shrubs, and rank vegetation. The reddish-brown larvae inhabit the burrows of wood-boring beetles and other insects, on whose larvae they feed. They pupate in cocoons in bark crevices, and the adults usually emerge all at one time in the spring, and fly by day, feeding on insects such as aphids.

3 Coniopteryx tineiformis. This tiny Lacewing, shown here at four times actual size, has a coating of a white, waxy secretion. It especially frequents oak and sallow in May and June and again in July and August. The pale yellow or pink larvae pupate under leaves in flat, round cocoons of white silk, in which the second generation overwinter.

4 Giant Lacewing (*Osmylus fulvicephalus*). The largest British Lacewing, with a 2-inch wingspan. It flies from May to August, mainly at night, with its slow, clumsy flight along woodland streams with dense vegetation. By day, like other Lacewings, it rests on the undersides of leaves. Though local, it is often common where found, especially in the south. The eggs are laid on streamside plants, and the larvae live amongst wet moss in which they hibernate. In spring, they resume feeding on the larvae of insects such as midges, and then pupate in yellowish, silken cocoons amongst the moss.

5 Sponge-fly (*Sisyra fuscata*). A small, semi-aquatic Lacewing which flies from May to September. The small, silk-covered clusters of eggs are laid on plants overhanging the water, into which the larvae drop and live as parasites on fresh-water sponges. When full-grown, they leave the water and hibernate in silken cocoons in bark crevices, and pupate there the following spring. Some larvae grow quickly and produce a second generation in late summer.

6 Brown Lacewing (*Kimminsia subnebulosa*). A family of small brown or greyish insects, of which this species is very common among deciduous trees and rank vegetation. In the south at least it produces several broods a year, and so can be found at all stages in any month, although it usually hibernates as a larva in a white, silken cocoon in moss or in a bark crevice. The eggs are laid, generally singly, under leaves, and the larvae are purple-brown.

7-8 Green Lacewings (family Chrysopidae). A family of mostly medium-sized green insects, though some are brown, with golden compound eyes. They usually fly in woods and hedgerows in two overlapping generations in late spring and summer, and usually hibernate as larvae in white, silken cocoons, pupating the following spring. They are most common in southern England and are not found in Ireland or Scotland. *Chrysopa septempunctata* (7) has a black spot between the antennae. When attacked, it produces an offensive-smelling secretion from stink glands on each side of the thorax. The females usually lay their greenish-yellow eggs in groups stuck to the undersides of leaves, and the first larvae to hatch devour the rest of the eggs. *Chrysopa perla* (8), which is heavily marked with black on head and body, deposits its eggs singly or in small groups under the leaves.

9 Snow Flea (*Boreus hyemalis*). The only British representative of the Moss-flies (Boreidae), a family of very small, ant-like, wingless insects. Instead of wings, the male Snow Flea has spine-like appendages curved back over its body. The female uses her long, sword-like ovipositor to plunge her eggs deep into the moss on which the larvae feed before pupating in the soil in September. The adults emerge in October, and are active by day throughout the winter, running and jumping rapidly over moss, earth, or even snow, being locally common in northern and eastern Britain.

10 Common Scorpion-fly (*Panorpa communis*). All three British species have beak-like faces, long antennae, and transparent wings marked with brown. The males have scorpion-like, but harmless, tails. This species, with strongly-marked wings, haunts sunny hedgerows, wood-edges, and rank vegetation from May to July. It takes short, rapid flights by day in search of its prey — usually dead insects or other carrion. The greyish-brown larvae, which also feed on carrion, hatch from batches of eggs laid in damp soil, burrow into the soil, and pupate in cells in their burrows.

LIFE SIZE (*No.* 3 × 4)

1 SIALIS LUTARIA 2 RAPHIDIA NOTATA

3 CONIOPTERYX TINEIFORMIS (× 4)

4 OSMYLUS FULVICEPHALUS 5 SISYRA FUSCATA

7 CHRYSOPA SEPTEMPUNCTATA 6 KIMMINSIA SUBNEBULOSA 9 BOREUS HYEMALIS

8 CHRYSOPA PERLA 10 PANORPA COMMUNIS

There are nearly 200 species of British Caddis-flies, most of which are nocturnal and come readily to artificial light, but a few (Nos. 5, 6, and 7, for example) fly by day. The sexes are in general alike. When at rest, most species fold their wings over their bodies in a roof-like attitude. The adults cannot feed, though some may sip liquids such as nectar. They have a complete metamorphosis. The eggs are normally laid in or near water in a mass, surrounded by a jelly-like substance. The larvae of all but *Enoicyla pusilla* (No. 6) are aquatic and breathe by gills. They have various methods of self-preservation and feeding. Some build and live in protective cases, made of secreted silk and sticks, stones, leaves, or shells, which they drag about the bottoms of ponds and streams; others are free-living. Some are vegetarian or omnivorous; others are strictly carnivorous. Species of *Hydropsyche* (No. 5) spin webs with which to catch their prey. The pupa is usually formed within the larval case, which is then attached to a firm support, and remains there until the adult emerges.

Fishermen have for generations used imitations of both adult and larval Caddis-flies as lures, and so some 'popular' names have grown up. But these are not always specific and sometimes refer to a group of unrelated species. Several families, for example, have been lumped together as 'Sedges'.

1 **Phryganea grandis.** This is 'the Great Red Sedge' of trout fishermen, and is one of the two biggest British Caddis-flies, a fairly powerful insect with a wingspan of nearly 2½ inches. It inhabits lakes and slow-moving, large rivers throughout the British Isles, and is most commonly on the wing from May to July. The larvae, which construct spiral-shaped cases of plant material, are omnivorous and even attack small fish.

2-3 **Limnophilidae.** There are many species in this family, which are variable and difficult to distinguish from each other. The adults of the two common species shown here fly from May to October by even quite small lakes and ponds, but often stay quiescent at the height of summer.
Limnophilus flavicornis (2) is one of the larger ones. The larvae make cases of a variety of materials, sometimes entirely of snail shells — even of a single species of snail. *L. vittatus* (3) is much smaller, and its transparent hindwings and rather pointed, often darkish forewings are distinct. The larvae make curved, cylindrical cases entirely of sand grains, usually to be found in standing water.

4 **Enoicyla pusilla.** The species of this genus differ from other Caddis-flies in that the adult female is wingless and the larvae are terrestrial, living at first in moss or grass at the foot of trees. They build cases like those of *Limnephilus vittatus*, but smaller. The smoky-grey adults appear in late autumn and early winter, and have been found only in Worcestershire.

5 **Hydropsyche angustipennis.** Little groups of these dark golden-brown and smoky-grey insects may be seen dancing in bright sunshine over waterside vegetation.

These are males; the females tend to sit amongst the foliage. They are on the wing from May to September, throughout the British Isles, often abundantly. The larvae, which inhabit running water, do not build cases, but catch their prey in webs spun near the shelter of stones or debris in which they live.

6 **Mystacides azurea.** A little Caddis-fly called 'Silverhorns' by fishermen, probably because of its very long, black-and-white antennae. Its long, narrow forewings give off a distinctive steel-blue reflection. Swarms of these insects fly close to the surface of either still or running water, during late afternoon, throughout most of the summer, over most of the British Isles, sometimes in large numbers. The larvae build tubular cases of stone fragments.

7 **Brachycentrus subnubilus.** Known as 'The Grannom' by fishermen, this Caddis-fly is on the wing as early as March and has usually disappeared by May. It is active in daylight, its smoky-grey wings being strongly marked with lighter streaks. The female is larger than the male and may often be seen carrying a bright-green ball of eggs beneath her body. The larvae live in flowing water, anchoring their cases to the weeds. It is widespread and sometimes locally abundant.

8 **Tinodes waeneri.** This rather nondescript insect is not easy to distinguish from several closely-related but less common species. Towards late evening, especially in July and August, the adults often sit in the foliage of trees near the pond or river where they breed. They are on the wing from May to September and are common over all the British Isles. The larvae inhabit mud tunnels on large stones or rocks, near the water's edge, sometimes in large colonies.

LIFE SIZE

1 PHRYGANEA GRANDIS   2 LIMNOPHILUS FLAVICORNIS
3 LIMNOPHILUS VITTATUS   4 ENOICYLA PUSILLA
5 HYDROPSYCHE ANGUSTIPENNIS   6 MYSTACIDES AZUREA
7 BRACHYCENTRUS SUBNUBILUS   8 TINODES WAENERI

This family, which includes the Heaths (p. 43), are, except for the Marbled White, all various shades of brown. Most are common or locally common, the Meadow Brown and Small Heath being very common. They have conspicuous eye-spots on the wings, which are usually black with white centres. The females are a little larger than the males and in most species rather paler, with larger blotches, and without the dark, oblique bands of scent glands on the forewings. Apart from the Grayling (p. 43) and the Wall Brown, they are not strong fliers. They overwinter, sometimes as pupae but usually as larvae, which feed on grasses even in winter in mild spells.

1 **Speckled Wood** (*Pararge aegeria*). A common butterfly in woods, wood borders, and shady lanes along hedgerows in the south and west of England, Wales, and Ireland, and locally on the west coast of Scotland; it is scarce or absent elsewhere. Its dappled appearance makes it almost invisible among the woodland leaves, which are also dappled by shafts of sunlight. The larvae hatch from greenish eggs, and those of the second generation overwinter either as larvae or more often as green pupae hanging suspended from the grass attached by a silken pad. In the spring the butterflies emerging from the hibernated pupae appear much earlier than those from the larvae, and consequently the second generation is out-of-step, giving the impression that there are more generations in a season than there really are.

2 **Wall Brown** (*Pararge megera*). An attractive tawny butterfly to be seen in sunny places in open country throughout Britain, except for northern Scotland. It can be mistaken for a medium-sized fritillary when seen in flight, but when it makes one of its frequent pauses to bask in the sunshine, it is easy to distinguish. It settles for a moment or so with wings spread wide on a sunny wall or other exposed place, and then flits quickly and restlessly on.
The first generation flies in May and June, and the second in July and August. The greenish eggs are laid on the grasses on which the larvae feed by night, even in winter during mild weather. The pupae are like those of the Speckled Wood.

3 **Marbled White** (*Melanargia galathea*). This butterfly, which looks very different from the rest of its family, flies in July and August in meadows with tall grasses and on rough hillsides and downs, in many parts of southern and central England and Wales; but it tends to be local. The white eggs are dropped at random on the ground among grasses on which the larvae feed. On hatching, the larva eats its empty egg-shell and then hibernates until early spring; then it feeds up and pupates on the ground in July. The butterfly emerges in two or three weeks from the creamy-yellow pupa, which is speckled with brown and suffused with pink. A completely black and another completely white

specimen were caught on one occasion, and were sold by auction in 1943 for £110.

4 **Hedge Brown** or **Gatekeeper** (*Maniola tithonus*). A common butterfly of the hedgerows and woodland rides and borders in southern England, south Wales, and locally in southern Ireland; further north it becomes progressively scarcer and more localized, and is absent from Scotland. It flies in July and August, and is attracted to bramble blossoms. The pale yellow eggs are laid on the grasses, on which the larvae feed at night before and after hibernation. They pupate in June. The pale, brownish-yellow pupae marked with dark-brown spots and streaks hang suspended from the grass, and the butterflies emerge in about three weeks.

5 **Meadow Brown** (*Maniola jurtina*). This very common butterfly flies from June to September in open country, especially rough grassland, and woodland clearings, almost everywhere in the British Isles. It is a lazy butterfly, reluctant to take to wing. It is drabber and larger than the Hedge Brown, and both sexes have variable amounts of orange-brown on the upperwings, the male being generally much darker.
The yellowish eggs are laid on grasses, and the larvae overwinter and pupate early the following summer. The green pupae, marked with dark-brown on the wing-cases, hang from grass stems or leaves. The butterflies emerge about four weeks later.

6 **Ringlet** (*Aphantopus hyperanthus*) is so named because of the ringed eye-spots. The male is sometimes almost black. Ringlets are about the same size as Meadow Browns and are often found in the same situations, but especially where the vegetation is lush. They fly in July and August and are common in most districts in the British Isles, except northern Scotland.
Like the Marbled White, the female scatters her eggs at random over the grass; these are pale yellow at first, becoming pale brown. The larvae hatch in August and feed by night, even during mild periods in winter, and in June they pupate on the ground. The butterflies emerge in about a fortnight from the pale yellowish-brown pupae marked with darker brown.

LIFE SIZE

1 SPECKLED WOOD   3 MARBLED WHITE   5 MEADOW BROWN
2 WALL BROWN   4 HEDGE BROWN   6 RINGLET

For a general note on the Browns see p. 40.

1 Mountain Ringlet (*Erebia epiphron*). This is the only true alpine butterfly found in Britain, to be seen in mountains of the Lake District in England above 1,800 feet and in the Grampians in Scotland above 1,500 feet. It has also been recorded from two Irish mountains. It flies only in the sunshine in June and July, and often in considerable numbers on boggy ground in its favoured haunts.

The Mountain Ringlet resembles the other northern species, the Scotch Argus, but is smaller and duller, with less distinct markings. The female deposits her eggs, yellow at first, later turning brown, on the mat grass. The young larvae feed by night until they are half-grown, then hibernate until the following March; by June they are full-fed and pupate in the grass tussocks, protected by a rudimentary silken cocoon. The stumpy pupa varies in colour from whitish-brown to pale green, marked with brown.

2 Scotch Argus (*Erebia aethiops*). This northern butterfly flies in the sunshine in July and August with a slow, fluttering flight, and favours rather damp grassland or sunny wood borders, but usually on lower ground than the Mountain Ringlet. It is widespread and common in Scotland, but has now disappeared from almost all its former haunts in northern England, and is not found at all in Ireland.

The female lays yellowish eggs which soon become purplish-brown. The young larvae hatch in September and feed by day on purple moor grass before hibernating through the winter; in spring they resume feeding, but only at night when full grown. They pupate in a silken tent on the ground in early July, and the butterflies emerge about a fortnight later. The pupae are yellowish-white with dark wing-cases and three brown lines along the back.

3 Grayling (*Eumenis semele*). The female of this fairly large species is more brightly coloured as well as larger than the male. Both sexes, when pursued, have the habit of alighting on the ground and displaying for an instant the bright eye-spots on the forewings. This diverts the attention of the pursuer from the vulnerable body to the comparatively unimportant wings. The butterfly then slides the conspicuous forewings behind the hindwings which so resemble the surroundings that it is difficult to see. The Grayling then leans over to an angle from the sun, which reduces the length of its shadow. It sometimes stays for long periods taking advantage of its camouflage.

The Grayling, which is a comparatively strong, fast flier, is on the wing from July to September. It frequents dry, often stony ground, such as chalk and limestone hills, heathland, and sand-dunes, and is locally common throughout the British Isles. The white eggs are deposited on grasses such as sheep's fescue. The larvae feed at night from September (even, at times, in the winter) until the following June, when they pupate in cocoons constructed underground — the only British butterfly to do this. The rich chestnut-brown pupa is smooth, plump, and rounded. The butterflies emerge after a month.

4 Large Heath (*Coenonympha tullia*). A locally plentiful butterfly from north Wales and Shropshire, northwards to the Shetland Isles, and throughout Ireland. Large Heaths vary considerably in colour from the north to the south of their range, and three distinct races are now recognized. The darkest with the largest and most conspicuous eye-spots is found farthest south (4a), and the lightest race in Scotland (4b). The female, as well as being paler, has more distinct eye-spots than the male.

Large Heaths fly in June and July in marshes and on damp, boggy moorland from sea-level up to 2,000 feet. Their habitat is often very windy, and they fly in a dogged manner near the ground. The greenish-yellow eggs are laid on white beak-sedge on which the larvae feed from July until October, when they hibernate for the winter. They resume feeding in March and pupate in May or June, the pupa, suspended from a grass stem, being bright green with brown streaks on the wing-cases. The butterfly emerges about three weeks later.

5 Small Heath (*Coenonympha pamphilus*). This little butterfly, smaller and generally brighter than the Large Heath, is probably as common as the Meadow Brown (p. 41), and even more catholic in its choice of habitat. It is found almost everywhere in the British Isles, except in the Orkneys, Shetlands, and Scilly Isles, from coastal marshes up to 2,000 feet in the mountains. It is perhaps most common in rough grassland, where it flits lazily among the grass blades.

There are two generations each year; the first is on the wing from May to July, and the second in August and September. The greenish eggs are laid on grasses such as annual meadow grass, on which the larvae feed in June and July, and as a second brood from September and in mild spells through the winter, until full-grown in April. The pupae are light green marked with brown and are suspended from grass blades and stems. The butterflies emerge after nearly a month.

LIFE SIZE

1 Mountain Ringlet   4 Large Heath

3 Grayling

2 Scotch Argus   5 Small Heath

43

FRITILLARIES (*Nymphalidae*)

The butterflies of the Nymphalid family are medium-sized to large, beautifully and often brightly coloured, and usually strong fliers (*see* also pp. 46, 48). The Fritillaries are a tribe of the Nymphalidae. The ones on this page lay their eggs on dog violet, on which the spiny larvae feed. The larvae of all species except the High Brown (4) hibernate through the winter and start feeding in the spring. They pupate suspended, head downwards, from a leaf, twig, or plant stalk. The female butterflies are a little larger and usually rather duller in colour than the males.

1 **Small Pearl-bordered Fritillary** (*Argynnis selene*). This species, which is easy to confuse with the Pearl-bordered, flies in June and July, occasionally later, in damp, open woodland and marshy country throughout Britain, especially in the north. It is not found in Ireland.

The larvae, when they hatch from the greenish-yellow eggs, hibernate amongst fallen leaves, and in early spring start feeding again, becoming fully-grown by mid-May. The pupae are brown, marked with black and metallic silver spots. The butterflies emerge after about two weeks.

2 **Pearl-bordered Fritillary** (*Argynnis euphrosyne*). This species is lighter coloured and has far fewer silver spots on the hind underwings than the Small Pearl-bordered. It also appears on the wing earlier, from early May to June, and is much more addicted to woodlands, haunting clearings and rides where wild flowers are abundant. It is widely distributed in Britain, but is much more scarce and local in the north, and in Ireland is known only in a few places in the west.

Its life history is much the same as that of the Small Pearl-bordered, except that the Pearl-bordered is about a month in advance, and the larvae and pupae differ slightly.

3 **Dark-green Fritillary** (*Argynnis aglaia*). This and the next two species have wing-spans of up to $2\frac{1}{2}$ inches. The Dark-green is so named because of the patches of dark-green scales on the yellowish-brown, silver-spotted, hind underwings. The female is more heavily marked with black than the male. It flies in July and August over open grassy country, such as downland, and also in open woodland in the north, and its fast, powerful flight makes it difficult to catch on the wing. Its range covers the whole of the British Isles, except East Anglia, and it is often common in suitable localities.

The yellowish eggs hatch in September, and the young larvae eat their empty egg-shells before hibernating for the winter. In April they begin to feed and pupate in June or July, the shiny, black and brown pupa being

sheltered in a tent of leaves bound together with silk. The butterflies emerge in about a month.

4 **High Brown Fritillary** (*Argynnis cydippe*). A rather paler Fritillary than the Dark-green, with underwings of a warmer buff colour, but lacking the patches of dark-green scales. The slightly larger female is paler than the male and has blunter wings. This is a typical butterfly of the woodlands, being rarely found outside them; it frequents the glades and rides where big thistles grow. It is locally plentiful throughout England south of the Lake District and flies in July and August. The pale greenish-yellow eggs overwinter and hatch early the following spring; and the larvae, like the Dark-green, pupate in June in the safety of a tent of leaves. The pupae, from which the butterflies emerge in about three weeks, are dark brown studded with two rows of brilliant metallic golden spines.

5 **Silver-washed Fritillary** (*Argynnis paphia*). The largest and most magnificent of the British Fritillaries, and another essentially woodland insect. It is found in most of the larger, older woodlands throughout the southern half of England and Wales, and all Ireland, but is very rare in northern England and Scotland. It is especially common in the New Forest, where it may be seen in open woodland rides and glades, from June to September, flying with effortless power and grace, and feeding from the bramble and thistle flowers.

It is easily distinguished from the other two large Fritillaries by the hindwings, which are streaked or 'washed' with silver on the underside, instead of being spotted. The female is larger and duller than the male and lacks the black streaks on the forewings. A greenish-silver form is not uncommon. The greenish eggs are laid singly in crevices in the bark of oak or other trees, and here, after eating their shells, the young larvae hibernate until next April, when they fall to the ground and make their way to dog-violet plants. When full grown in June, they change into light, yellowish-brown pupae studded with brilliant, metallic gold points, and the butterflies emerge about three weeks later.

LIFE SIZE

1 SMALL PEARL-BORDERED FRITILLARY   2 PEARL-BORDERED FRITILLARY
3 DARK-GREEN FRITILLARY   4 HIGH BROWN FRITILLARY
5 SILVER-WASHED FRITILLARY

For a general note on Fritillaries see p. 44.

1 Marsh Fritillary (*Euphydryas aurinia*). A medium-sized Fritillary with a much weaker flight than the two Pearl-bordered species (p. 45). It varies both in size and colour-pattern, the Irish race being brighter than the British one. It is to be found in colonies in marshy areas and damp meadows, though also on hill-tops, and is locally common throughout the British Isles, though more scarce in the eastern counties. It flies in May and June.

The bright yellow eggs are laid in clusters on devil's-bit scabious, and soon become dark. The larvae hatch in July and live gregariously in a silken web which they spin around the foodplant. In the autumn, they spin a stronger web in which they hibernate through the winter, and continue to use as a base when they resume feeding in March or April. When fully grown, they pupate suspended in a leafy tent, as the Dark-green Fritillary does (p. 44). The dumpy pupa is pale brown and black, spotted with orange.

2 Glanville Fritillary (*Melitaea cinxia*). Although recently introduced into one locality in Hampshire, the home of this butterfly is the Isle of Wight, where it flies commonly enough in May and June on the rough cliff slopes of the south coast. It is very like the Heath Fritillary, which is not found on the island, but it has much paler, whiter, frillier-looking underwings. The female lays her yellowish eggs in large batches on sea or ribwort plantain, on which the larvae feed gregariously within a silken web from July onwards, and then hibernate in the web. They come out of the web to feed in the following March, but still keep together in a colony. In April or May they pupate, the dark-brownish pupae having black and orange marks and orange points.

3 Heath Fritillary (*Melitaea athalia*). A very local woodland insect, found plentifully only in a few woods in Kent, Essex, Sussex, Devon, and Cornwall, where its foodplant, common cow-wheat, flourishes. In June and July it flies, rather like the Glanville Fritillary, in a weak, fluttering manner. The female deposits her yellowish eggs in large batches on the cow-wheat in June and July, and the larvae hatch in August and feed gregariously in a silken web. They pupate about a month later than the Glanville Fritillaries do. The small, plump pupa is yellowish-white marked with black and orange.

4 Duke of Burgundy Fritillary (*Hamearis lucina*). In spite of its name, this butterfly belongs to a different family, the Riodinidae, of which it is the only European representative, and is more closely related to the Lycaenidae

(p. 52) than the Nymphalidae. It is locally plentiful in the south of England, but rare in the north and absent from Scotland and Ireland. It flies in May and June, flitting rapidly in sunny woodland rides and clearings carpeted with wild flowers. The female, which is slightly larger and paler than the male, deposits her yellowish eggs on cowslip or primrose leaves, on which the woodlouse-like larvae feed from June to August. The dumpy, hairy little pupa, brownish-white in colour and spotted with black, is attached with silk to the underside of the leaf of the foodplant, where it remains for about ten months before the butterfly emerges.

5 White Admiral (*Limenitis camilla*). 'Admiral' is a corruption of 'Admirable'. This handsome butterfly belongs to the same family as the Fritillaries, and is found quite commonly in larger woods in southern England, but not elsewhere. It flies gracefully and effortlessly in the open rides and clearings around the bramble blossoms in July and August. Its blackish-brown and white upperwings contrast with the orange-brown underwings, spotted and banded with white. Some specimens are almost entirely black. The female, which is slightly larger than the male, deposits her shiny green eggs on the more scraggy honeysuckle plants in the shade, and the larvae hatch about a week later, and when not feeding, lie along the midribs of a leaf. In the autumn, each larva constructs a winter shelter by pulling the edges of a honeysuckle leaf together and attaching it firmly with silk to the plant stem. In the spring they resume feeding until June, when they pupate, hanging head downwards from leaves or plant-stems. The pupa, from which the butterfly emerges about a fortnight later, has a glistening silver patch on a green background.

6 Purple Emperor (*Apatura iris*). One of the largest and most magnificent of British butterflies, with a lovely, iridescent, purple sheen showing in certain lights in the male's upperwings, though not in those of the larger, browner female. It is rather rare and extremely local, being confined to the larger oak-woods of the southern half of England, where the male soars around the tree-tops on its powerful wings in July and August, rarely descending to the ground, except when attracted to carrion or a bright light. The female flies lower in search of sallow bushes on which she lays her greenish-olive eggs, which become black just before hatching in August. The slug-like larva develops horns after the first moult and, while still small, hibernates on a sallow twig until the following March. It pupates in June hanging head downwards from a sallow leaf. The pupa, from which the butterfly emerges about a fortnight later, is horned and greenish-white, resembling a curled sallow leaf.

LIFE SIZE

1 Marsh Fritillary   2 Glanville Fritillary   3 Heath Fritillary
4 Duke of Burgundy Fritillary   5 White Admiral   6 Purple Emperor

VANESSIDS (*Nymphalidae*)

All Vanessids are brightly coloured and most frequent gardens, attracted by buddleia, Michaelmas daisies, and other flowers. The sexes differ little, though the females are slightly the larger. They are strong fliers. Some migrate to southern Europe in the autumn, the next generation returning to breed in May or June. Other species hibernate successfully in Britain and appear early in the spring, mate, and lay their green eggs on the foodplant. The pupae hang head downwards from the foodplant or nearby and are usually greyish-brown studded with metallic gold or silver spots — hence the name 'chrysalis' meaning golden.

1 **Red Admiral** (*Vanessa atalanta*). A brilliant butterfly often to be seen in gardens, even in cities, and round over-ripe fruit in orchards. As it is an immigrant, the numbers fluctuate from year to year: in some years it is found over most of the British Isles, though more in the south. Each spring, immigrants arrive from southern Europe and lay their eggs singly on stinging nettles. The larvae hatch from June onwards and feed inside leafy tents spun together with silk. They pupate in July or August fastened inside leafy shelters, and the butterflies emerge a fortnight later. Many fly south in the autumn, though some hibernate, usually unsuccessfully.

2 **Painted Lady** (*Vanessa cardui*). Like the Red Admiral, this lovely migrant comes from around the Mediterranean and reaches the southern shores of Britain in May and June, sometimes earlier. Many more come in some years than others, and in good years they spread over the British Isles, attracted even to town gardens by buddleia and other flowers. The females lay their eggs singly on thistles or nettles, and the larvae feed within silken webs among the leaves. They pupate inside leafy tents spun together with silk. The butterflies emerge in about a fortnight and fly from July to October, when many migrate south. The rest soon die.

3 **Small Tortoiseshell** (*Aglais urticae*). A very common British butterfly, reinforced by immigration from the Continent. It differs from the Large Tortoiseshell not only in size but also in the heavy black scaling around the body. Two generations occur each year, in June, and again in August and September, and there is occasionally a third. From October the butterflies hibernate, often in buildings, reappearing in early spring to pair. The females lay their eggs in clusters on stinging nettles, and the tiny black larvae feed and grow together in silken webs among the leaves. When over an inch long and dark greyish green with yellow stripes, they separate and pupate, the butterflies emerging about two weeks later.

4 **Large Tortoiseshell** (*Nymphalis polychloros*). A rare butterfly in Britain, being most often seen in wooded areas of East Anglia and Kent in July and August or, after hibernation, in April and May, when it pairs. The female deposits her large batches of yellowish-green eggs on the topmost twigs of elms and occasionally other trees, where the larvae feed together under the shelter of a web which they spin along a branch. When full grown in July, they drop to the ground and pupate in the undergrowth. The butterflies emerge in about two weeks.

5 **Peacock** (*Nymphalis io*). A large, handsome butterfly, with four conspicuous eye-spots and a powerful flight. It is found throughout the British Isles, most commonly in the south. Like the Tortoiseshells, it passes the winter in the butterfly stage, sleeping in a hollow tree, building, or similar place, and reappears in early spring, when it pairs. The olive-green eggs are laid in May in large clusters on stinging nettles, where the black, spiny larvae feed and grow together in a silken web until July, when they separate and pupate in the undergrowth. The pupa is usually grey-green or brown with two pointed horns on the head. The butterflies emerge in July and are on the wing until they hibernate.

Camberwell Beauty (*Nymphalis antiopa*). A large purple-brown butterfly with a yellowish border to its wings, called the Mourning Cloak butterfly in America and Scandinavia. It is not known to breed in Britain, but a few reach the east coast from Scandinavia in the autumn, perhaps on timber boats, and occasionally survive the winter. It was first seen in Britain at Camberwell, south-east London, in 1748 — hence its English name.

6 **Comma** (*Polygonia c-album*). This tattered-looking butterfly has white comma-like marks on the underside of the hindwings. It is most common in southern England and Wales, especially around the Severn; it is rarer further north and unknown in Scotland and Ireland. It frequents wooded districts as well as parks and gardens. There are two generations each year, flying between July and October. The first Commas to appear are much paler and brighter than the later ones, with yellowish-brown underwings. The second generation, which are darker, fly until October, when they hibernate, clinging with closed wings to branches and looking like dead leaves. In early spring they awake and pair. The female lays her eggs, often singly, on hops or stinging-nettles, and the larvae feed singly in small webs under the leaves until they pupate in June. The spangled pupa is usually pinkish-brown. The butterfly emerges in about ten days.

LIFE SIZE

1 RED ADMIRAL   2 PAINTED LADY
3 SMALL TORTOISESHELL   4 LARGE TORTOISESHELL
5 PEACOCK   6 COMMA

For a general note on the Lycaenidae see p. 52.

1 Holly Blue (*Celastrina argiolus*). A woodland butterfly, also seen in parks and gardens, generally common in southern England, Wales, and Ireland, very local in northern England, and absent from Scotland. Its pale-blue underwings sparsely spotted with black and lilac-blue upperwings distinguish it from the other Blues on page 53. The female has more black on the wing borders than the male.
Except in northern Ireland, there are two generations each year. The larvae of the spring brood hatch in June from greenish eggs laid on holly, and the butter-flies emerge in late summer; those of the second brood feed on ivy and pass the winter as pale-brown pupae attached to the leaves by silk threads. The summer butterflies are darker than the spring ones, especially the females.

2 Small Copper (*Lycaena phlaeas*). A very active little butterfly, common in open country throughout the British Isles, except for some of the Scottish islands. It could only be confused with the female of the Large Copper. Both species are characterized by their iridescent copper-orange colour. There are three generations in a year between April and October. The female, which is like the male, lays her yellowish eggs singly on the undersides of sorrel or dock leaves, on which the larvae feed, those of the third generation passing the winter in a silken shelter on the foodplant. The brown pupa, from which the butterfly emerges after a month, is attached to the foodplant by a silken girdle.

3 Large Copper (*Lycaena dispar*). The British race became extinct over 100 years ago, but in 1927 the very similar Dutch race was introduced into Wood Walton Fen in Huntingdonshire, where the larva feeds on the great water-dock and is flourishing under protection. Although the larva has no honey-gland, it does secrete a sweet substance and so is attended and protected by ants (*see* p. 52).

4 Green Hairstreak (*Callophrys rubi*). The only British butterfly with really green wings (on the underside only). Like the other Hairstreaks, the sexes are much alike, and there are thin, hair-like marks on the underwings, from which comes the name Hairstreak. The Green Hairstreak is common throughout the British Isles, flying in May and June in bushy places where broom, gorse, dyer's greenweed, bilberry (in Scotland), dogwood, and its various other foodplants grow. The larvae, which hatch from greenish eggs laid singly, are cannibals. They pupate in August in the leaf-litter on the ground, and the butterflies emerge from the brown pupae next spring.

5 Brown Hairstreak (*Thecla betulae*). The female is larger and prettier than the male, having patches of orange on her forewings. The tawny undersides of both sexes distinguish them from other Hairstreaks. They are rare in Ireland but locally plentiful in southern England and Wales, where they frequent wood borders and overgrown hedgerows, flying in August and September. The white eggs are deposited singly on blackthorn and hatch the following spring; the larvae pupate under-neath the leaves of the foodplant, and the butterflies emerge about three weeks later from the brown pupae, marked with darker brown. This, and other Hair-streaks, except the green, have slender tails on each hindwing which, like the eye-spots of the Browns, deflect the attention of an enemy from the vulnerable head.

6 Purple Hairstreak (*Thecla quercus*). The blackish wings shot with purple iridescence are brighter in the female than in the male, even though the iridescence is confined to patches on the forewings. Purple Hair-streaks are locally plentiful in oakwoods throughout the British Isles, except northern Scotland and Ulster, where they fly in July and August high around the oaks, on which the females lay their white eggs. These hatch in the spring, and the larvae descend from the trees to pupate on the ground in June, when huge numbers are eaten. The butterflies emerge a month later from the reddish-brown pupae.

7 White-letter Hairstreak (*Strymonidia w-album*). This butterfly has white hair-lines on the underside of the hindwing, which look like a letter 'W'. The female has longer tails than the male but lacks the small patch of scent-scales on the forewings which all male Hair-streaks carry. They are locally plentiful in woods and hedgerows over most of southern England and Wales in July and August, flitting rapidly around their food-plants, common or wych elms. They also visit bramble and privet blossoms. The larvae hatch from the green, turning brown, eggs in early spring and pupate in late June. The light-brown pupae with dark wing-cases are attached to the leaves or twigs by a silken girdle.

8 Black Hairstreak (*Strymonidia pruni*). A very rare butterfly in Britain, known only in a few places in the Midlands, where it flies, in June and July, in woodland clearings and borders and blackthorn thickets. It is distinguished from the White-letter by orange blotches (more numerous in the female) on the wing margins of the upperside and by having no 'W' marks. The larvae hatch in spring from brown eggs laid singly on the highest spikes of old blackthorn bushes the previous July; they pupate in June, attached by silk threads to the twigs or leaves. The black and white pupa, from which the butterfly emerges in about eighteen days, looks exactly like a bird dropping.

1♂

1♀

1♂

1♂

6♀

6♂

6♂

4♀

4♂

3♂

5♂

7♀

7♂

7♂

3♂

3♀

5♀

5♂

2♂

8♂

8♀

2♀

LIFE SIZE

1 HOLLY BLUE 2 SMALL COPPER 3 LARGE COPPER
4 GREEN HAIRSTREAK 5 BROWN HAIRSTREAK 6 PURPLE HAIRSTREAK
7 WHITE-LETTER HAIRSTREAK 8 BLACK HAIRSTREAK

51

The Blues, Coppers, and Hairstreaks (p. 51) are a family of small, swift-flying, and often brightly-coloured butterflies. The Blues, except for the Common Blue, belong mainly to southern England. The males are mostly some shade of blue and the females brown. The larvae, shaped like woodlice, have a gland in the back which secretes a sweet fluid, and are 'milked' and in some cases 'farmed' by ants (*see* No. 7). The pupae are stout, smooth, and rounded. Of the eleven British species, the Long-tailed, the Short-tailed, and the Mazarine are rare migrants.

1   **Small Blue** (*Cupido minimus*). The tiny female is sooty brown in colour, but the male has a powdering of pale blue scales. They are locally common in southern England and Ireland, in dry, sunny, grassy places, especially chalk and limestone hills and near the sea, where their foodplant, kidney vetch, is plentiful.
The females lay light-green eggs on the flower buds of kidney vetch. The larvae hibernate fully grown in frail cocoons spun on the foodplant and closely resembling the seed-pods. In May they change into pinkish-grey pupae, spotted with black, and the butterflies emerge about three weeks later.

2   **Silver-studded Blue** (*Plebejus argus*). The distinctly purple male differs from the Common Blue in having black borders to the upperwings. Both sexes have black spots with metallic silver-blue centres or 'studs' on the margins of the underside of the hindwings. This Blue, found commonly on dry heathland in the south and east and locally on chalk downs and sandy coasts, flies from mid-June to August. The white eggs are laid on heather or leguminous plants and hatch the following April. The larvae pupate about mid-summer, the greenish-yellow pupae hanging from the foodplant by silken threads. The butterflies emerge some three weeks later.

3   **Brown Argus** (*Aricia agestis*). The only member of the group to have no blue colouring in either sex. The marginal band of red-orange blotches is larger in the female, which differs from the female Silver-studded Blue in having no silver-blue 'studs' on the underwings. A Scottish race, which has smaller red-orange blotches, white instead of black spots on the forewings, and no black centres to the white spots on the underwings, is thought in fact to belong to the continental species, *A. allous*.
The Brown Argus is generally common on southern chalk and limestone hills, sandy areas, and coastal marshes, but is absent from Ireland. In the north the single generation usually flies in July and August; in the south there are two — in early and in late summer. The larvae, which hatch from white eggs, feed on rock-rose and hemlock stork's-bill, and the second generation hibernates and feeds again in the spring. The yellow-green pupae, from which the butterflies emerge in two weeks, are suspended among the leaves of the foodplant.

4   **Common Blue** (*Polyommatus icarus*). The commonest of the Blues is found in grassy places all over the British Isles, except for the outer Scottish islands. They gather in the evenings in long grasses where they rest head downwards. Except in the extreme north, there are two generations between May and September. The greenish eggs are usually laid on bird's-foot trefoil or rest-harrow; the larvae of the second brood hibernate at the base of the foodplant and pupate in April. The butterflies emerge from the mainly green pupae in about two weeks.

5   **Chalk-hill Blue** (*Lysandra coridon*). These Blues are locally common on chalk and limestone hills in the south and east, and fly in July and August. The dingy-brown females are slightly larger than the Adonis Blue females and usually have some pale silvery-blue scales and fewer, duller orange spots on the uppersides. Their greenish eggs, laid on horse-shoe vetch, hatch the following April, and the larvae pupate in June in holes in the soil, forming light greenish-brown pupae. The butterfly emerges after a month.

6   **Adonis Blue** (*Lysandra bellargus*). The silver-blue of the male is distinct from the more violet Common Blue, and the female has blue scales and brighter orange spots on the upperside, and a rather chequered appearance of the white wing fringes. Its habitat and foodplant are the same as the Chalk-hill, but it is more local and is not found in East Anglia. It flies in May and June, and again in August and September. The young larvae of the second generation hibernate after hatching and in the spring form yellowish-brown pupae in crevices in the soil.

7   **Large Blue** (*Maculinea arion*). A rare butterfly of a few rough grassy localities in Cornwall, Devon, Somerset, and Gloucestershire, and the largest of the Blues. The female has heavier black spots on the upperwings than the male. It flies in June and July.
It has a remarkable life-history associated with ants. After hatching from a bluish-white egg, its larva feeds on wild thyme flowers until after the third moult, when it leaves the foodplant and wanders about. Before long it is found by an ant of the genus *Myrmica* (p. 150), which strokes the gland on its back and drinks the sweet droplets which are secreted. Eventually the larva hunches its back, and this stimulates the ant to carry it off to its nest. Here it remains through the winter, feeding on the ant larvae and growing white and grublike. In late spring it pupates, and three weeks later the butterfly emerges from the yellow-brown pupa, walks out of the nest, expands and dries its wings, and flies off.

LIFE SIZE

1 SMALL BLUE 2 SILVER-STUDDED BLUE 3 BROWN ARGUS
4 COMMON BLUE 5 CHALK-HILL BLUE 6 ADONIS BLUE
7 LARGE BLUE

53

WHITES (*Pieridae*)
AND SWALLOWTAIL (*Papilionidae*)

The Pierid family are mostly common, medium-sized, slow-flying butterflies, with bottle-shaped, ribbed eggs. They include the Whites and the Yellows (p. 57). The very common Large and Small Whites are the well-known 'Cabbage Whites', serious pests of cabbages and allied crops. In some years there are particularly large immigrations from the Continent.

1 **Large White** (*Pieris brassicae*). There are two generations each year, in the spring and again in late summer, those of the second generation being more heavily marked with black on the upperside and yellower below. The female, which has black spots on the forewings, lays her large batches of yellow eggs on cruciferous plants and nasturtiums, and the evil-smelling, showy larvae feed gregariously until nearly full-grown. They would be an even more serious vegetable pest were it not that many die of a bacterial disease and others are parasitized by an ichneumon fly (p. 146). The larvae pupate on walls, fences, and tree trunks, the greenish-grey, black-spotted pupae being attached by a silken pad and girdle. They overwinter as pupae and the butterflies emerge in April.

2 **Small White** (*Pieris rapae*). This even more common butterfly than the Large White is less heavily marked with black, especially in the first generation. The male has one and the female two black spots on each forewing. The first generation appears in April, the second in August, and in some years there is a third. The light-yellow eggs are laid singly on cruciferous plants, and the solitary, inconspicuous larva leaves the foodplant when full grown and pupates in the same manner as the Large White. The pupae vary in colour to match their surroundings, and those of the second brood overwinter.

3 **Green-veined White** (*Pieris napi*). A common butterfly of open country, especially marshy meadows. It is more heavily marked with black scales than the Small White, especially in the second generation. On the underside these are intermingled with yellowish scales and are concentrated along the veins, which produce the 'green-veined' appearance. The female is darker and more heavily spotted with black than the male. They fly from April to June, and again in August and September. The greenish eggs are deposited singly on hedge garlic, charlock, and other Cruciferae, and the larvae pupate, usually on the foodplant; the second generation overwinters as pupae. These are usually green, but vary a good deal according to their background.

4 **Bath White** (*Pontia daplidice*). A very rare and irregular immigrant to the south coast from southern Europe. It resembles superficially the Green-veined White, but has a large, square, black spot in the centre of each forewing. The larvae usually feed on mignonette and in 1945, when an exceptional number reached Britain, they also fed on hedge mustard.

5 **Orange-tip** (*Anthocharis cardamines*). The female lacks the male's gay patches of orange on the forewings, and at rest she differs from the Small or Green-veined White by her dappled-green underwings. In May and June Orange-tips fly in a rapid, fluttering manner through meadows and along hedgerows and wood borders over almost all the British Isles except north Scotland. The greenish-white eggs, laid singly on the flower heads of hedge garlic, lady's smock, and charlock, soon turn orange, then dark violet. The larvae feed from June to August on the seed-pods, which they closely resemble, and sometimes on each other. They pupate attached head upwards to the foodplant by a pad and girdle of silk, and as the winter proceeds, like the seed-pods, they gradually turn brownish.

6 **Wood White** (*Leptidea sinapis*). A delicate, weak-flying butterfly, the only British member of its group, most of which belong to the New World. It is found very locally in woodland rides and clearings in southern England, especially the West Country and Monmouthshire. There is an Irish race with distinctly greenish undersides.

Wood Whites fly in May and June and, in the southwest, again in late July, when they are paler than the earlier insects. The males have blacker markings than the females. The larvae, which hatch from yellowish eggs, feed on tufted vetch, bird's-foot trefoil, and other legumes in June, and again in August. They pass the winter in beautiful green pupae attached to the foodplant.

7 **Swallowtail** (*Papilio machaon*). This is the only British species of the beautiful Papilionid family. It is now confined to the Norfolk Broads, where it is still plentiful, and attempts are being made to re-introduce it in Wicken Fen, Cambridgeshire. Occasionally, a few Swallowtails of the paler continental race reach southern England and temporarily establish themselves, feeding on carrot.

British Swallowtails fly in a strong, flapping manner in May and June. The females, which are a little larger than the males, deposit their yellow eggs singly on milk parsley, a plant found only in fen country. The full-grown larva has a strong-smelling, orange, forked organ behind the head, which it shoots rapidly in and out when alarmed. It pupates in August, attached head upwards to a reed-stem by its tail and a silken girdle. Some butterflies emerge in August and lay eggs which produce larvae in September, but most overwinter as green or brownish pupae.

LIFE SIZE

1 LARGE WHITE   2 SMALL WHITE   3 GREEN-VEINED WHITE
    4 BATH WHITE   5 ORANGE-TIP   6 WOOD WHITE
        7 SWALLOWTAIL

YELLOWS (*Pieridae*)
AND SKIPPERS (*Hesperiidae*)

For a general note on the *Pieridae* see p. 54 and on Skippers see p. 58.

1 **Pale Clouded Yellow** (*Colias hyale*). Both this and the Clouded Yellow are immigrants to Britain from the Mediterranean region, seldom penetrating further north than the southern counties. The Pale Clouded Yellow normally arrives from May onwards and seeks out lucerne and clover fields in open country, but it is usually rare. The females deposit their yellowish eggs (which later turn orange-red) on these plants, on which the larvae feed. The greenish-yellow pupae are attached to the foodplant by a silken girdle, looking like withered leaves. A new generation of butterflies emerges in the autumn, but larvae or pupae from these rarely survive the winter. Their appearance in Britain is, therefore, dependent on migration.

The Pale Clouded Yellow is a strong, rapid flyer. The female, whose ground colour is nearly white, can be confused with the variety *helice* of the Clouded Yellow, but it is much less heavily marked with black on the hindwings.

2 **Clouded Yellow** (*Colias croceus*). Small numbers of this swift and strong-flying immigrant to the British Isles from southern Europe reach the southern counties in most years, and occasionally a great number arrive, producing a large generation of butterflies in the late summer and autumn. When this happens, the Clouded Yellows spread further northwards, some penetrating into Scotland and Ireland. The first arrivals appear in May or June, and the females lay their eggs (at first light yellow, then turning orange) on lucerne, clover, and other legumes, in open country. The larvae feed in June and July, and then pupate on the foodplant and produce butterflies in August. Their offspring feed during the autumn, but are unable to hibernate and are killed off in the winter. The pupae are like those of the Pale Clouded Yellow. The broader wings of the female have yellow spots in the dark borders. The fairly frequent female variety *helice* has a white or pale-yellow ground colour instead of orange-yellow.

3 **Brimstone** (*Gonepteryx rhamni*). The fairly large, sulphur-yellow males are among the first butterflies to appear in the spring because they hibernate through the winter in the butterfly state, hidden in evergreen bushes and looking very much like pale leaves. They appear on the wing early in spring, followed later by the females, which are greenish-white and could be mistaken for a 'Cabbage White' when flying, but they have a stronger, more direct way of flying. After pairing, the females deposit their greenish eggs, which later turn yellow, singly on buckthorn leaves. The green larvae, which harmonize splendidly with the leaves, have a habit of resting along the midrib of a leaf. They hatch in June and pupate in July, attached horizontally underneath a leaf or twig by a silken pad and girdle. The pupae are green and extremely leaf-like. The butterflies emerge after about a fortnight and in the autumn seek a place to hibernate. On fine days a few may still be flying about as late as early November.

Although typically a woodland butterfly, the Brimstone often frequents hedgerows away from woods, and sometimes gardens throughout most of England and Wales, and south and west Ireland, where it is generally fairly common.

4 **Dingy Skipper** (*Erynnis tages*). A dingy, rather moth-like butterfly found over most of Britain but commonest in the southern half of England and Wales, and very local in Ireland. The sexes are alike, except that the males are rather darker. They fly in May and June, darting rapidly from flower to flower in open, rough, grassy places where wild flowers are abundant. The yellow eggs are laid singly on bird's-foot trefoil, on which the larva feeds inside a leafy tent, spun together with silk, until August when, fully fed, it constructs another leafy shelter in which to hibernate for the winter. In April it pupates inside this shelter. The narrow pupa, from which the butterfly emerges after about four weeks, is reddish-brown with greenish wing-cases.

5 **Grizzled Skipper** (*Pyrgus malvae*). An unmistakable little butterfly with its chequered black and white upper-wings and grey, brown, and white underside. It frequents much the same haunts as the Dingy Skipper, but also flies in woodland clearings. The male is rather lighter and greyer than the female. It is commonest in the south and is not found north of Yorkshire nor in Ireland. It flies from late April to June, darting from flower to flower and feeding with horizontally spread wings. The greenish-white eggs are usually laid singly on wild strawberry leaves, or sometimes bramble or raspberry. The larva feeds from June until August within a shelter formed by drawing a leaf together with silk. It pupates at the base of the foodplant within a cocoon formed from leaves and silk, from which the butterfly emerges the following spring. The pupa is light brown with greenish wing-cases.

LIFE SIZE

1 PALE CLOUDED YELLOW   2 CLOUDED YELLOW   3 BRIMSTONE
4 DINGY SKIPPER   5 GRIZZLED SKIPPER

SKIPPERS (*Hesperiidae*)

There are eight British species of this large family (*see* also p. 57), many of whom are South American. They are mostly small, and named because of their rapid, darting flight. The family is primitive and very moth-like in some ways, for instance, the stout, often hairy bodies and prominent eyes. Some species, the Dingy Skipper (p. 57), for example, often rest with their wings folded flat over the body, as moths do, but most hold the hindwings horizontally and the forewings tilted at an angle of 45°. The males, but not the females, have an oblique black streak of scent scales on the forewings.

1 **Chequered Skipper** (*Carterocephalus palaemon*). This distinctive little Skipper, with its chequered livery, is rather like the frailer Duke of Burgundy Fritillary (p. 47). It inhabits grassy woodland glades and clearings in a few places in the east Midlands of England and Inverness-shire, Scotland, and flies in late May and June. The female, which is slightly larger and paler than the male, lays yellowish eggs singly on certain brome grasses. The larvae draw together the edges of a grass blade to form shelters, from which they emerge to feed from June until October. They hibernate in similar tubes formed from two grass blades, and in April spin several blades together and pupate inside them. The narrow pupae, from which the butterflies emerge about six weeks later, are straw-coloured, with brown streaks on the wing-cases.

2 **Small Skipper** (*Thymelicus sylvestris*). This differs from the Essex Skipper, in having orange instead of black tips to the undersides of the antennae, orange-brown rather than golden-brown upper-wings, and a more pronounced oblique black streak of scent scales on the male's forewings.

Small Skippers fly in July and August in rough, grassy places in both open country and woodland clearings where flowers are plentiful. They are widespread and common in southern England, very local in Wales, and absent from Scotland and Ireland. The females deposit the yellowish eggs in rows of three to five inside a sheath of Yorkshire fog or other grass. The larvae hatch in August and, having eaten the egg-shells, spin silken cocoons inside the grass sheath and hibernate until the spring. They construct feeding and pupating shelters, as does the Chequered Skipper, and pupate in June or July. The butterflies emerge about two weeks later from the thin, mainly light-green pupae.

3 **Essex Skipper** (*Thymelicus lineola*). This is much like the Small Skipper, and the two butterflies often fly together in July and August; but the Essex is largely confined to the low-lying Thames Estuary country, where it is locally abundant as far west as the eastern suburbs of London. It is also found, but less commonly, in the eastern counties of England, and has been seen in Devon and Somerset. It inhabits rough grassland, marsh pasture, salt marshes, and also sometimes chalk downs, and its habits, foodplants, and life history are like those of the Small Skipper except that they overwinter as eggs.

4 **Lulworth Skipper** (*Thymelicus acteon*). This butterfly resembles the Small and Essex Skippers in life history and habits but is duller in colour, and the female has a more conspicuous orange-yellow semi-circular mark on each forewing. It is entirely restricted to the Dorset coast (especially in the Isle of Purbeck around Lulworth Cove) and to the neighbouring coast of Devon. Here it flies quite abundantly over the sea cliffs and steep hillsides in July and August and sits in the sun with its wings held in the typical Skipper manner. The pearl-white eggs are laid in long rows of usually five to fifteen together, and the usual food-plants are slender brome grass and couch grass.

5 **Silver-spotted Skipper** (*Hesperia comma*). A butterfly found only on the chalk and limestone hills of south-east England, where it is locally common and to be seen sitting on the flowers of the dwarf or stemless thistle in the characteristic Skipper pose. It resembles the Large Skipper, but the undersides of its hindwings are greener and spotted with silver-white. The female is rather darker.

It flies actively in August. The white eggs (which turn yellow later) are usually laid singly on sheep's fescue or other grass, and hatch the following spring. The larvae construct feeding shelters by spinning together grass blades. In July they construct grassy cocoons near the ground, the mainly olive-brown pupae being furnished with hooks to grip the interior of the cocoon. The butterflies emerge after about ten days.

6 **Large Skipper** (*Ochlodes venata*). Though like the Silver-spotted Skipper, it has no silver-white spots on the underwings. The female is slightly larger and darker. It is common in open country and woodland rides and clearings over most of England, Wales, and southern Scotland, but rare or absent elsewhere. It is on the wing from June to August, and suns itself in the usual Skipper fashion. The creamy-white eggs, laid singly on grasses, soon become yellow. The larva, on hatching, constructs a tube-like shelter from a grass blade, from which it feeds and in which it hibernates until the spring. After feeding up, it pupates in May in a cocoon spun together with grass and silk, the long, mainly-brown pupa being secured by hooks. The butterfly emerges some three weeks later.

LIFE SIZE

1 CHEQUERED SKIPPER   2 SMALL SKIPPER
3 ESSEX SKIPPER   4 LULWORTH SKIPPER
5 SILVER-SPOTTED SKIPPER   6 LARGE SKIPPER

HAWK-MOTHS (*Sphingidae*) 1.

The majority of the Sphingidae or Hawk-moths (*see* also pp. 63, 65) are large, with long, rather narrow wings. The males of most species are rather smaller and in some species darker than the females. They are all swift, powerful fliers, and some are well-known migrants. The eggs are green, in some species brighter green than in others, and the larvae of most species are large and have characteristic horns at their rear ends. Most Hawk-moths pupate in cocoons or chambers a few inches or more underground, and the pupae are generally dark brown. Of the 900 or so species of Hawk-moths in the world 18 occur, though only 9 breed, in Britain.

1 Death's-head Hawk-moth (*Acherontia atropos*). A large and spectacular, but harmless, moth, the bulkiest of the British species and easily recognized by the mask or skull-like pattern on its back, combined with its huge size and distinctive colouring. Another remarkable feature is its ability to squeak when alarmed; it produces the sound by forcing air through its short, thick proboscis. It sometimes enters old-fashioned skep-type bee-hives in search of honey, and it has been said that the squeaking pacifies the bees. But, in fact, judging by the moth's behaviour when disturbed, the function of the squeak is probably to frighten off enemies.

These Hawk-moths cannot live through a British winter, so their presence in Britain depends entirely on immigration from the Continent. Like other migrant insects, the numbers vary considerably from year to year, but they are usually rare. Most arrive in late spring and the autumn. In a favourable year there may be two generations, arrivals in May giving rise to larvae in July, and the resulting moths producing more larvae in September and moths in late October and November. Death's-head Hawk-moths have been seen in many parts of the British Isles, but most often in the south-east of England.

They usually lay their eggs on potato plants, but also on woody nightshade, jasmine, and some other plants. The larvae are almost 5 inches long when full grown and are usually yellowish, although brownish and blackish forms are sometimes found. The large dark-brown pupae and also the larvae are discovered in some years by workers digging in potato fields.

2 Convolvulus Hawk-moth (*Herse convolvuli*). This powerful, fast-flying moth has an even greater wing-span than the Death's-head, but a smaller body. Its larger size and grey rather than brown wings distinguish it from the Privet Hawk-moth. Its remark-ably long proboscis, about 5 inches long, enables it, while hovering in flight, to probe plants with tubular flowers, such as petunias, for their nectar. Convolvulus Hawk-moths fly from dusk until a late hour. Like many other migrants, the numbers reaching Britain vary considerably from year to year, and in some years they are common, though they are usually scarce. They fly direct to the British Isles from the Mediterranean region, and so are naturally more frequent in southern England, though they do occasionally reach the extreme north and west. They are most usually to be seen at dusk in gardens and parks with luxuriant flower borders. In some years, the first arrivals appear as early as June, but August and September are the more usual months for big immigrations. They do not usually breed in Britain, but occasionally the females lay their bright-green eggs on convolvulus, usually the field bindweed, on which the large, very variable, usually green or blackish-brown, larvae feed. They pupate deep in the soil, and the large reddish-brown pupa is easily recognized by the way in which the proboscis case is curved round like a jug handle.

3 Privet Hawk-moth (*Sphinx ligustri*). Another very large, powerful moth, but smaller than the Convolvulus and browner, with usually pinker hindwings. The male is both smaller and has thicker antennae than the female. Although it has been seen in many places, it is common only in the southern counties of England, and is not found in Ireland. It flies swiftly at night from late May to July in open country, frequently visiting town gardens, even in London, in search of flowers. The very large larvae feed on privet and also lilac, from July to September, and for such brightly coloured creatures, they are not easy to see. They burrow several inches into the soil to pupate and the large, chestnut or chocolate-brown pupa remains underground throughout the winter.

LIFE SIZE

1 DEATH'S-HEAD HAWK-MOTH
2 CONVOLVULUS HAWK-MOTH
3 PRIVET HAWK-MOTH

61

For a general note on Hawk-moths see p. 60.

1 Lime Hawk-moth (*Mimas tiliae*). A medium-sized Hawk-moth which varies a good deal in the colour and pattern of the wings. The dark blotches in the centre of the forewings sometimes join up to form a band, are sometimes reduced to one, and occasionally are almost or completely absent. The male is usually greener than the female and has a slimmer body. Like several members of the family, the Lime Hawk has a characteristic way of resting with the forewings held apart, exposing part of the hindwings and the whole of the abdomen; this gives it a triangular appearance resembling a bunch of leaves.

Lime Hawk-moths are widespread in Britain south of Yorkshire, but are common only in the southern counties, especially in the London area. They fly after dark in May and June where their foodplants, lime and elm trees, are common, and they are often to be seen in town parks and gardens. The eggs are laid singly or in pairs on the foodplants, on which the larvae feed in July and August. The larvae pupate just beneath the surface near the foot of a lime or elm tree, and the moth emerges the following spring.

2 Eyed Hawk-moth (*Smerinthus ocellata*). These moths may be seen sitting during the daytime on tree-trunks or fences, with their wings held in the same manner as the Lime Hawk-moth. If disturbed, they suddenly disclose their pink hindwings, revealing two large and striking blue eye-spots, which have the function of frightening away predators. Eyed Hawk-moths are found throughout England, Wales, and southern Scotland. They are generally common in the southern half of England, widespread and fairly common in Ireland, but more local and scarce further north. In June they fly at night where their usual foodplants, sallow, willow, and apple, are plentiful. The females lay their yellowish-green eggs singly on the leaves of these trees, on which the larvae feed from July to August. Occasionally, in favourable years, there is a partial second generation in late July. They overwinter as shiny dark-brown pupae, buried a little way under the surface of the soil.

3 Poplar Hawk-moth (*Laothoë populi*). The commonest British Hawk-moth, to be found almost everywhere in the British Isles, flying after dark in open country from May to August. There are often two generations in a year. The eggs are laid, usually singly, on the leaves of poplars, willows, and sallows, on which the larvae feed from June to September. Moths of the second generation may appear from late July. The larvae pupate just below ground near or under the foodplants, and the moths overwinter as pupae.

4 Narrow-bordered Bee Hawk-moth (*Hemaris tityus*). This and the Broad-bordered are day-flying moths which mimic bumble bees and so gain protection from predators. The resemblance to bumble bees is striking when the moth hovers over the blossoms of bugle, lousewort, or other favoured flower; but when it flies off, its exceedingly rapid, darting flight is quite unlike that of a bumble bee. The wings of both species, when they emerge from the pupae, are coated with loose scales, most of which are lost during the first flight, leaving the wings largely transparent, like a bee's. The two species are difficult to separate in flight, but at rest, the narrower, blackish borders of the wings of the Narrow-bordered are distinctive.

The moths fly in May and June, usually in the mornings only, in bogs, fens, and open places in or near woods where the foodplant, devil's-bit scabious, grows. They are locally plentiful over most of the British Isles. The eggs are deposited under the leaves of the foodplant, on which the larvae feed in June and July. The insects overwinter in the pupa stage on the ground inside coarse, silken cocoons, which are covered with earth and debris.

5 Broad-bordered Bee Hawk-moth (*Hemaris fuciformis*). This moth is distinguished from the Narrow-bordered by the much broader, reddish-brown borders of the wings and the tail tuft, which is buff with blackish sides. It flies in May and June, usually in the morning, in the neighbourhood of woods, especially where bugle and rhododendrons grow. It is widespread and locally common in England and Wales, but is not found north of Yorkshire nor in Ireland. The bright-green eggs are laid singly, and the larvae feed on honeysuckle from June to August. They pupate in the same way as the Narrow-bordered, and the moths emerge the following spring.

6 Humming-bird Hawk-moth (*Macroglossum stellatarum*). When seen hovering in the sunshine in front of a flower, with rapidly-beating wings and long 'tongue' extended, this handsome moth very closely resembles a humming bird, and is different from any other British moth, except perhaps the Bee Hawks.

The Humming-Bird Hawk is an immigrant to the British Isles from southern Europe, usually arriving from June onwards. The females lay their eggs on lady's bedstraw and closely-related plants, on which the larvae feed in July and August. On the Continent there are two or three generations each year. The larvae change to light-brown pupae in weak, silken cocoons, and the moths emerge three weeks later. The adult moths have been seen in every month of the year, all over the British Isles, though in the south far more often from June to September. In some years they are abundant; in others rare. They frequent parks and gardens well provided with flowers. Occasionally a moth may hibernate successfully through the winter, but most either migrate south or perish.

LIFE SIZE

1 LIME HAWK-MOTH   2 EYED HAWK-MOTH
3 POPLAR HAWK-MOTH   4 NARROW-BORDERED BEE HAWK-MOTH
5 BROAD-BORDERED BEE HAWK-MOTH   6 HUMMING-BIRD HAWK-MOTH

For a general note on Hawk-moths (*Sphingidae*) see p. 60.

1 Elephant Hawk-moth (*Deilephila elpenor*). A beautiful insect which owes its name to the tapering shape of the larva, which is reminiscent of an elephant's trunk. Elephant Hawk-moths fly at night in June and hover over flowers, probing them for nectar with their long probosces. They are found over most of England, Wales, Ireland, and southern Scotland, wherever their chief foodplants, rose-bay, great willowherb, and marsh bedstraw, grow abundantly, though they are rather rare in the north. They come to rose-bay willowherb in some towns and cities, and were especially common on London bomb sites in the years soon after the Second World War.

The larvae, which hatch from light-green eggs, feed by night from July to September. Though normally blackish-brown, a green form is not uncommon. They display their large eye-spots if molested. The insects pass the winter as light-brown pupae, marked with darker brown, lying on the ground in coarse cocoons of earth and plant debris, interwoven with silk.

2 Small Elephant Hawk-moth (*Deilephila porcellus*). This smaller Elephant Hawk flies at night in May and June in southern England, and later in the north. It freely visits flowers, especially honeysuckle, and is locally common in Britain (except north-west Scotland) and Ireland. The green eggs are usually laid singly on bedstraws, especially lady's bedstraw, and sometimes willowherbs, on which the larvae feed between June and September. Although very like the last species, the larvae lack the typical Hawk-moth horn, having instead only a stump. They pass the winter in the pupal stage, both pupa and its cocoon being similar to the Elephant Hawk's.

3 Emperor Moth (*Saturnia pavonia*). A fairly large insect which is the only British representative of the famous silk-moth family *Saturniidae*, to which belong some of the world's largest moths. In August the Emperor spins an elaborate coarse-brown silk cocoon (of no commercial value) on the foodplant, usually heather or bramble. The cocoon is rather pointed at one end, inside which is a ring of stout, converging spines. It is through this end that the moth emerges from the blackish-brown pupa next spring. The spines open outwards when pressed from inside, but thwart any enemy trying to enter from outside.

The large eye-spots on all four wings of both sexes prevent confusion with any other British moth. The female is much larger than the male, and is a lighter grey, lacking his tawny hindwings. She flies only at night, but the males fly very fast by day in April and May over heaths, moors, marshes, and similar open country when the sun shines. They seek out the un-mated females, which sit in the herbage, attracting them with their scent. Emperor Moths are locally common throughout the British Isles, and their colourful larvae, which are remarkably inconspicuous

in the heather, hatch from olive-brown eggs and feed from May to August.

4 Puss Moth (*Cerura vinula*). A large, handsome member of the Prominent family of moths (p. 66), its thick body suggesting a soft, fluffy cat. The female is larger than the male and has darker hindwings. In May and June the moths fly commonly almost everywhere in the British Isles where their foodplants, sallows, willows, and poplars, grow. The reddish-brown, almost hemi-spherical eggs are laid, one to three together, on the upper surface of the leaves, and the larvae hatch in June. When full-grown, the larva is a remarkable creature; the curiously angled, white-bordered, purple markings on its otherwise bright-green body break up its general outline so that it is difficult to see among the leaves of its foodplant. If it is discovered by an enemy it rears up into a fearsome attitude, as shown in the picture, draws in its head, exposing a bright red rim, and lashes out with two long, red threads which protrude from its forked tail. If the attacker is still not deterred, the larva squirts a jet of strong formic acid from a gland in its thorax. In August or September, the larva selects a crevice in the bark of the tree on which it has fed, spins a silken cocoon mixed with chewed-up pieces of wood which, when it hardens, makes it virtually indistinguishable from its surroundings, and changes inside this into a dark, reddish-brown pupa. The following spring the moth softens the cocoon with a secretion of caustic potash and cuts its way out with a specially modified part of the pupa-case, which remains on its head until it has emerged.

5 Lobster Moth (*Stauropus fagi*). In general appearance this member of the Prominent family (p. 66) is like a dark greyish-brown or blackish Puss Moth. The male is smaller than the female and has thicker antennae. The brown larvae, which hatch from creamy-white eggs, are even more fantastic than the Puss larvae, somewhat resembling a lobster. Two of the three pairs of 'true' legs are unusually long, and their tail-ends, which they hold up over their backs, are swollen and bear two stiff filaments. When the larva is alarmed, it rears up, extends these long forelegs, and rapidly vibrates the second pair, mimicking the ferocity of a spider. If this does not frighten off the intruder, it squirts formic acid, as the Puss Moth does. The larvae feed from July to September, chiefly on beech, but also birch, oak, and hazel. They spin strong silken cocoons between dead leaves, in which they pupate, and the moths usually emerge the following May or June. In some years there is a partial second generation. The Lobster Moth is very locally distributed in woods, mainly beechwoods, throughout southern and eastern England south of Yorkshire. It is very rare in Wales and Ireland, apart from Co. Kerry, and unknown in Scotland.

LIFE SIZE

1 ELEPHANT HAWK-MOTH   2 SMALL ELEPHANT HAWK-MOTH
3 EMPEROR MOTH
4 PUSS MOTH   5 LOBSTER MOTH

PROMINENTS (*Notodontidae*)

There are twenty-five British species of Prominents, nine of which are shown in this book (*see* also p. 64). They are all swift, night-flying moths, which come readily to bright lights. Most species have a tuft of scales sticking out from the centre of the hind margin of each forewing, which are particularly prominent (hence the name) when the moth is at rest.

1 **Swallow Prominent** (*Pheosia tremula*). Both this and the very similar Lesser Swallow Prominent (*Ph. gnoma*) are generally common over most of the British Isles where poplars grow, except in the far north of Scotland; the Lesser Swallow is the more common in Ireland. The white, wedge-shaped mark at the angle between the outer and inner margins of the forewings is wider and much more distinct in the Lesser Swallow. Swallow Prominents are on the wing in May and again in August. The females, which are slightly larger and have thinner antennae than the males, lay their creamy-white eggs under poplar and, occasionally, sallow leaves, and the larvae of the first generation feed in June and July. They pupate in rough silken cocoons among the leaves of the foodplant. Those of the second generation spin their cocoons underground, where they remain until the moths emerge next May. The pupae are reddish brown.

2 **Pebble Prominent** (*Notodonta ziczac*). A common moth almost everywhere in the British Isles where its foodplants, willows and sallows, grow, especially low-lying areas. Two generations of the moths fly each year: in May and June, and again in August. The females, which have browner hindwings than the males, deposit their greenish eggs in little groups on the leaves of the foodplants, and the larvae feed in June and July and again in August and September. They pupate in rough cocoons just below ground under the food-tree, those of the second generation remaining throughout the winter as reddish-brown pupae.

3 **Coxcomb Prominent** (*Lophopteryx capucina*). This is the commonest British Prominent, occurring almost everywhere in the British Isles, especially in wooded districts. In the south there are two generations, in May and June, and again in July and August. This rather variably coloured moth resembles the scarce Maple Prominent (*Lophopteryx cucullina*), but the latter has largish pale patches on the margins of the forewings. The females, which have thinner antennae than the males, deposit their whitish eggs in small groups under the leaves of beech, birch, hazel, oak, sallow, and occasionally other trees, and the larvae feed from June to October. They pupate on the ground near the foodplant in silken cocoons covered with earth, and those of the second generation pass the winter in this stage. The pupae are dark brown.

4 **Pale Prominent** (*Pterostoma palpina*). The male, in particular, has a long body; the female is similar but paler and with thinner antennae. Although found over most of England, Wales, southern Scotland, and Ireland, Pale Prominents are common only in the south, where two generations fly each year: in May and June, and again in July and August. The larvae hatch from whitish eggs, usually laid in little groups under leaves of poplars, or sometimes willows and sallows; they feed between June and October, and then form silken cocoons interwoven with pieces of earth amongst the debris under the food-tree, and transform into dark-brown pupae. Those of the second generation overwinter until the following spring.

5 **Buff-tip** (*Phalera bucephala*). A large, handsome moth which, when at rest, looks just like a freshly-broken, lichen-covered twig. Both sexes are alike, except that the female is the larger. Buff-tips fly in June and July, and are common almost everywhere in the British Isles where trees are plentiful, especially in the south, and even in London. The female lays her whitish eggs in batches under the leaves of elm, hazel, lime, oak, and sallow. The larvae feed gregariously in the open from July to September. Their bright black-and-yellow colour and offensive smell warn enemies that they are distasteful. When nearly full-grown they separate and, when ready to pupate, descend the tree and construct underground chambers, in which they pass the winter as blackish-brown pupae.

6 **Small Chocolate-tip** (*Clostera pigra*). This is the commonest and smallest of the three British Chocolate-tips, the other two, *C. curtula* and *C. anachoreta*, having more conspicuous chocolate tips to the forewings. Like the Buff-tip, the Small Chocolate-tip resembles a short, broken stick when resting by day. There are two or three generations each year: in May, July, and often again in the autumn, and they are generally plentiful over most of the British Isles, especially in boggy or marshy districts where the foodplants are abundant. The female, which has a stouter body than the male, deposits her light-olive-green eggs more or less in short rows under the leaves of creeping willow, or occasionally sallows. The larvae feed at night from June to September, and live during the day in shelters formed by spinning together the terminal leaves of the foodplant. When about to pupate, they spin together a leaf and change into reddish-brown pupae within silken cocoons.

LIFE SIZE

1 SWALLOW PROMINENT   2 PEBBLE PROMINENT
3 COXCOMB PROMINENT   4 PALE PROMINENT
5 BUFF-TIP   6 SMALL CHOCOLATE-TIP

THYATIRIDS, HOOK-TIPS (*Drepanidae*), AND TUSSOCKS (*Lymantriidae*)

1 **Buff Arches** (*Habrosyne pyritoides*). This and the next two species belong to a family of rather small night-flying moths, the Thyatiridae. The Buff Arches flies in June and July, and occasionally also in the autumn, in well-wooded districts over most of England and Ireland, but is common only in the south. It is scarce in Wales and virtually unknown in Scotland. The female lays her eggs on bramble leaves, on which the larvae feed at night from July to October. By day they hide in leafy shelters drawn together with silk. They pass the winter as blackish pupae in underground pupal cells.

2 **Peach-blossom** (*Thyatira batis*). A locally plentiful moth in wooded districts over much of the British Isles, though scarce in Scotland. It flies in May and June, and sometimes again in the autumn. The greenish eggs are laid in groups along the edges of bramble leaves, and the larvae feed between June and September. Some feed up rapidly, pupate in late July, and produce moths in the autumn; but most pupate in September, producing moths the following year. They spin silken cocoons among the bramble leaves and change into light-brown pupae, marked with darker brown.

3 **Yellow Horned** (*Achyla flavicornis*). A very early moth which is on the wing by March and April in birch woods over most of the British Isles, although rather local in Ireland. The larvae hatch in May from eggs laid in small groups on birch twigs, on the leaves of which they feed by night until they pupate in late July; by day, each larva hides in a leaf drawn together with silk. The reddish-brown pupae lie in silken cocoons among the leaf litter under the foodplant, where they remain through the winter.

4 **Pebble Hook-tip** (*Drepana falcataria*). The members of this family have hook-tipped wings, and the queer, knobbly larvae have no claspers, or gripping legs, at the tail-end. The Pebble Hook-tip normally flies at night in May and June, and again in August. When at rest, it resembles a curled, dead leaf. It is common where there are birches throughout the British Isles (the Scottish race is whiter than the typical form), except in Ireland, where it is widespread, but uncommon. The female, which lacks the feather-like antennae of the male, lays yellowish eggs in little rows on the undersides of birch leaves, on which the larvae feed in June and July, and again in the autumn. The pupating larva curls over a leaf, inside which it spins a silk cocoon; the pupa is dark brown.

5 **Chinese Character** (*Cilix glaucata*). A distinctive little moth which lacks the characteristic hook-tips of the family; its curious name refers to the markings on the wings. It avoids its enemies by looking like a bird-dropping when it is resting during the daytime. The female, which is slightly larger than the male, deposits her dark, shiny eggs on blackthorn or hawthorn leaves, on which the larvae feed in June and July, and again in the autumn. The larvae pupate in brown silk cocoons among leaves or under the bark of the foodplant; and the brown pupae have grey wing-cases. The moths emerge the following May and June and fly by night; a second generation appears in July and August. They are generally common in hedgerows and wooded districts throughout England, Wales, and south Scotland, but less so in Ireland.

6 **Vapourer** (*Orgyia antiqua*). A very common moth of the Tussock family to be seen everywhere, even in towns and cities, over most of the British Isles, except Ireland where it is uncommon. The Tussocks are a very hairy group of moths, even the pupae being hairy. The larvae have four conspicuous brushes of yellow hairs, with longer hairs on the head and last segment, and these can cause an unpleasant skin rash to anyone handling them at any stage. The wingless female, which is quite unlike the male, remains with her cocoon, waiting for a male to find her. After mating, she lays a large batch of brownish-white eggs over the empty cocoon, which hatch the following spring. The larvae feed on various plants, especially oak, lime, and hawthorn, and in June construct silken cocoons mixed with their own hair, usually in the bark crevices of trees. The moths emerge from the shiny black pupae between July and September. The male usually flies by day.

7 **Pale Tussock** (*Dasychira pudibunda*). The female is larger and paler than the male, and her antennae are not feather-like. She lays her large batch of eggs around a twig or plant stem. The contrasting colours of the larvae, like those of the other Tussocks, serve to warn enemies that they are distasteful to eat. They feed between June and October on hops and various trees, and spin strong double-shelled cocoons of silk among the leaves. The moths emerge from the brown pupae the following May or June and fly by night. They are commonest in the south in wooded districts, are locally common in south Ireland, and extremely rare in Scotland.

8 **Yellow-tail or Gold-tail** (*Euproctis similis*). Its golden-yellow tail tuft distinguishes this moth from the similar, but rarer, Brown-tail (*E. chrysorrhoea*). The male is smaller than the female and has a black dot on each forewing. They fly by night in June and July and are common in most wooded districts throughout England and Wales, but extremely rare in Scotland and Ireland. The female usually lays her egg-batches on hawthorn leaves, covering them with protective hair from her tail tuft. In August the brightly-coloured larvae hatch and hibernate through the winter while still young; they pupate the following June on the foodplant in cocoons of silk mixed with their own hairs, which are irritating to predators.

LIFE SIZE

1 BUFF ARCHES   2 PEACH-BLOSSOM MOTH
3 YELLOW HORNED MOTH   4 PEBBLE HOOK-TIP
5 CHINESE CHARACTER   6 VAPOURER
7 PALE TUSSOCK   8 YELLOW-TAIL MOTH

The Noctuidae are a huge family of small or medium-sized moths, sometimes called Owlets. There are more than 6,000 species in the world and some 300 in Britain, of which only the commonest are described here. The Noctuidae usually have fully-developed probosces, with which they can suck nectar from flowers. The males and females of most species are alike. They almost all fly at night only and are often attracted to artificial light. By day they usually rest on tree-trunks, walls, or in dark places, with their hindwings folded beneath the forewings. The larvae of most species feed by night, and the pupae, which are usually some shade of brown, are most often formed in cells underground.

1   **Turnip Moth** (*Agrotis segetum*). This, like all the species on this page and Nos. 1 – 6 on p. 72, belongs to the sub-family Noctuinae, the majority of which have rather dull camouflaged forewings and grey or white hindwings. This very common moth is a serious pest of root-crops such as turnips, in the young stems and tap-roots of which the larvae feed from July to October. The white eggs are laid in batches on or near the stems of the foodplant. Most of the larvae hibernate full-fed through the winter and pupate underground in the spring. The male of this very variable moth, which flies in June and often again in the autumn, usually has paler forewings and whiter hindwings than the female.

2   **Heart and Dart** (*Agrotis exclamationis*). A very common moth with black markings on its rather variable forewings and a distinctive black mark on the front of the thorax. The female, which has darker hindwings than the male, lays her batches of purplish-marked, white eggs on the leaves of plants, including turnips and other vegetables. The larvae hatch in the summer, complete their growth the following spring, and then pupate in underground cells. The moths fly in June and July, and a partial second generation may appear in the autumn in favourable years.

3   **True Lover's Knot** (*Lycophotia varia*). A pretty little moth, generally common on heather moors and heaths, which flies from June to August by night, and occasionally also by day. Both sexes have the white-marked, reddish forewings. The larvae hatch from batches of whitish eggs, feed at night on heather and heaths, and hibernate in the winter. The pupae lie in thin silken cocoons on the ground beneath the foodplant.

4   **Purple Clay** (*Diarsia brunnea*). The rather variable forewings are usually some shade of purplish or reddish-brown. The moths fly in June and July in wooded districts, and the larvae, which hatch from batches of darkish eggs, feed on plants such as wood-rushes, docks, brambles, and sallows, until they hibernate for the winter; in the spring they often feed on young birches before pupating in thin silken cocoons just below ground.

5   **Flame Shoulder** (*Ochropleura plecta*). The markings on the forewings of this moth make it easy to identify when open, but when resting by day among dead leaves on the ground, it is difficult to see. It flies in May and June, and, as a second generation, in August and September. The larvae feed on plants such as bedstraw and cultivated vegetables in June and July, and again in the autumn, and the second generation overwinter as pupae in crevices in the soil.

6   **Setaceous Hebrew Character** (*Amathes c-nigrum*). The name of this moth comes from the usually pinkish and black marks on the forewings. It flies from May to early July and again from August to October, the second generation being much commoner than the first, possibly because of immigrants from the Continent. Little is known about its life history, in spite of its abundance, though it is thought that the larvae, which hatch from batches of pale, shiny eggs, feed on chickweed, groundsel, and similar plants from autumn to spring. There are, presumably, also larvae in the summer from moths of the first generation.

7   **Square-spot Rustic** (*Amathes xanthographa*). The forewings of this very common moth vary from whitish-brown, through red and greyish-brown, to blackish, and even the characteristic pale, square-shaped marks vary. The moths fly in August and September, and the larvae, which hatch from batches of pale eggs ringed with darker colour, feed on grasses and many other plants before hibernating for the winter. They feed again in the spring, and in May prepare underground pupal cells, but do not pupate until July.

8   **The Flame** (*Axylia putris*). A generally common moth, except in the extreme north, which resembles a dead piece of wood when resting during the day. It flies at night in June and July, sometimes later. The larvae feed from July to October on a wide variety of plants, including hedge bedstraw and stinging-nettle, and the dark-brown pupae overwinter in pupal cells in the soil.

LIFE SIZE

1 Turnip Moth   2 Heart and Dart
3 True Lover's Knot   4 Purple Clay
5 Flame Shoulder   6 Setaceous Hebrew Character
7 Square-spot Rustic   8 The Flame

For a general note on Noctuidae see p. 70.

1 Green Arches (*Anaplectoides prasina*). When fresh, the forewings of this moth are a lovely deep green, but later they fade to yellowish. It flies in woodlands at night in June and July and is locally plentiful over most of the British Isles, especially in south-east England, but it is rather scarce in Scotland. The larvae, which hatch from batches of dark-coloured eggs, feed on docks, plantains, bramble, and other plants from July until they hibernate for the winter; in the spring they attack the fresh shoots and leaves of sallow or bilberry. They pupate in April or May in an underground chamber.

2 Lesser Yellow Underwing (*Euschesis comes*). This common moth could be confused with the Lunar Yellow Underwing (*E. orbona*), but the latter has a noticeable black spot at the tip of each forewing. The forewings of both sexes vary a great deal in their shade of brown and the conspicuousness of their other markings. Like the other Yellow Underwings on this page, this species baffles its enemies by the way it shows its colours. When resting amongst the ground vegetation its brown forewings make it inconspicuous; but on being disturbed it flies off in a rapid, erratic manner, suddenly revealing its yellow hindwings, until after a relatively short flight it plunges into the vegetation again, closes its wings, and apparently disappears. This moth flies in July and August, usually at night. The female lays dark-coloured eggs in batches on various grasses and other plants, on which the larvae feed from August until they hibernate. In the spring, they resume feeding, often climbing on to the young leaves of hawthorn and sallows. They change into brown pupae on the ground beneath the foodplant in May.

3 Large Yellow Underwing (*Noctua pronuba*). This moth uses the same protective devices as the last species, but even more effectively owing to its flashier yellow hindwings, larger size, and generally more striking appearance. The forewing colours range from whitish-brown to dark brown, those of the stouter females being generally darker than the males. The moth has a long season, being on the wing from June to October, though most abundantly in July and August. The white eggs are laid in dense batches on various plants, and the larvae feed by night from August right through the winter. In the spring they change into chestnut-brown pupae in underground pupal cells. During the day they hide in the soil. Larvae from the earliest moths often produce moths in the autumn of the same year.

4 Broad-bordered Yellow Underwing (*Lampra fimbriata*). A very showy moth with rich orange-yellow hindwings with wide black borders. The forewings vary from pale brown to dark brown, those of the stouter females usually being the lightest. The Broad-bordered, which is on the wing from June to August, is locally common in wooded districts throughout the British Isles, except the extreme north. The larvae feed at night on birch, hazel, sallow, hawthorn, and other trees until they hibernate for the winter. They resume feeding on the young leaves in the spring. They hide by day on the ground beneath the foodplant and climb up again in the evening. In May they pupate underground, the pupae being a dark chestnut-brown.

5 Red Chestnut (*Cerastis rubricosa*). This is rather like a small Purple Clay (p. 70), but its forewings are always reddish-purple suffused with grey. It flies in early spring, often visiting sallow catkins at dusk, and is more or less plentiful almost everywhere in the British Isles. The yellowish-white eggs are laid in batches on plants such as dandelion, dock, and groundsel, on which the larvae feed from April to June. The larvae pupate just below ground, and the stout, dark-brown pupae remain there until the moths emerge in the spring.

6 The Gothic (*Naenia typica*). The pale, criss-cross markings on the forewings look rather like an old spider's web. The moths fly at night in June and July in gardens, hedgerows, and along wood-borders, and are locally plentiful. The shiny white eggs are laid in batches on a great variety of plants and bushes, and the larvae feed gregariously from August until they hibernate. In the spring they feed again, but separately, and in May they pupate in cells just below ground. The pupae are a glossy, chestnut-brown.

7 Beautiful Yellow Underwing (*Anarta myrtilli*). This species is common on heather moors and heaths, where, when resting, its forewings closely match its surroundings. When disturbed, it displays the same protective device as the Lesser Yellow Underwing. It flies actively by day between April and August, and the female lays her eggs on heather and heaths, on which the larvae feed from June to October. The brown pupae overwinter in silken cocoons mixed with soil particles on the ground beneath the foodplant.

LIFE SIZE

1 GREEN ARCHES   2 LESSER YELLOW UNDERWING
3 LARGE YELLOW UNDERWING   4 BROAD-BORDERED YELLOW UNDERWING
5 RED CHESTNUT   6 THE GOTHIC
7 BEAUTIFUL YELLOW UNDERWING

For a general note on Noctuidae see p. 70.

1 **Cabbage Moth** (*Mamestra brassicae*). A very common, rather drab, brownish moth with a whitish spot on each forewing. It flies in June and July and occasionally a partial second generation flies in the autumn. The female deposits her eggs in large batches on a great many plants, on which the larvae feed from July to October, when they are often a pest of cabbages and other vegetables, especially in gardens. The chestnut-brown pupae overwinter in underground cells and are easily found in the garden.

2 **Dot Moth** (*Melanchra persicariae*). Another familiar garden species, at least in the south, which is easily recognized by the conspicuous white spot on each forewing. The moths fly in July and August, but are only really common in the south, and are absent from Scotland. The larvae feed from August to October on many different wild and cultivated plants and then pupate, and the dark-brown pupae lie underground throughout the winter.

3 **Bright-line Brown-eye** (*Diataraxia oleracea*). This pest of tomatoes can be recognized by the yellowish spot and white line near the edge of each forewing. The forewings vary from reddish to dark brown. The moths normally fly from May to July, and the female lays her batches of pale eggs on many plants besides tomatoes, including stinging nettle, bindweed, and goosefoots, on which the larvae feed from June to September. The reddish-brown pupae pass the winter underground.

4 **Broom Moth** (*Ceramica pisi*). The forewings of this moth vary from greyish brown to reddish brown, those of the Scottish race being purplish brown, and there is a yellowish line near the edge of each forewing. The striking larvae, which hatch from batches of dark-coloured eggs, feed from July to September on broom, bracken, and many other plants, and are often to be seen in the open during daylight. The dark, reddish-brown pupae lie below ground until the moths emerge the following May, June, or July.

5 **The Campion** (*Hadena rivularis*). A usually common moth of open country which is on the wing in June and July, and again as a partial second generation in August, and visits flowers freely after dark. The female deposits her batches of dark-coloured eggs on bladder campion and related plants, and the larvae feed on the foliage and developing seeds from July to September. Some feed-up quickly and produce moths in August.

The remainder pupate in the autumn in silken cocoons on the ground, the moths emerging the following summer from the brown pupae.

6 **Hebrew Character** (*Orthosia gothica*). The English name refers to the distinctive black mark on each forewing which is said to resemble a Hebrew letter. This is a very common moth of early spring, which visits sallow catkins at dusk and is attracted to light in wooded districts throughout the British Isles from March to early June. The female lays her batches of white eggs dotted and ringed with grey on various low plants, trees, and bushes, on which the larvae feed from April to June, and then pupate in the soil. The brown pupae overwinter.

7 **Small Quaker** (*Orthosia cruda*). Dingy, undistinguished little moths, usually with a dusky spot on each forewing. Like the last species, they are fond of sucking nectar from sallow catkins at dusk. They fly in March and April and are more or less common in wooded districts, except for the extreme north. In Ireland they are rather local. The larvae, which hatch from batches of whitish eggs marked with brown, feed from April to June on trees such as oak, sallow, aspen, and hawthorn. The brown pupae overwinter in the soil at the base of the foodplants.

8 **Common Quaker** (*Orthosia stabilis*). A rather larger moth than the Small Quaker and, instead of the dingy spots, it has noticeable pale circular marks on the forewings which vary from light greyish brown to dark brown. Common Quakers visit sallow blossoms, and the females deposit dark-coloured eggs in batches on oak, sallow, birch, and other trees, on which the larvae feed from April to July. The brown pupae remain amongst the debris underneath the foodplant through the winter, and the moths emerge the following March or April.

9 **Clouded Drab** (*Orthosia incerta*). A well-named species for, although its forewings vary tremendously from specimen to specimen, they are usually some drab shade of brown, grey, or black, clouded with a darker colour. The female deposits her brown-banded, yellowish eggs in batches on sallow, oak, and various other trees and bushes, on which the larvae feed from April to July, and then pupate underground. The moths emerge from the reddish-brown pupae the following March and April, when they are familiar visitors to sallow catkins.

LIFE SIZE

1 CABBAGE MOTH   2 DOT MOTH
3 BRIGHT-LINE BROWN-EYE   4 BROOM MOTH
6 HEBREW CHARACTER   5 THE CAMPION
7 SMALL QUAKER   8 COMMON QUAKER   9 CLOUDED DRAB

For a general note on Noctuidae see p. 70.

1 Pine Beauty (*Panolis flammea*). The reddish-marked forewings of this moth closely match the bark of Scots pine-trees on which it sits during the day. After dark from March to May it flies in or near pinewoods, generally plentifully throughout Britain, except in the extreme north; in Ireland, though widespread, it is scarce. It frequently feeds at sallow catkins. The larvae, which hatch from dark-coloured, shiny eggs, harmonize extremely well with the leaf-needles of the Scots Pine on which they feed from May to July. They pupate within flimsy silken cocoons spun up in crevices of pine trunks or amongst moss, and the reddish-brown pupae overwinter.

2 Feathered Gothic (*Tholera popularis*). Though rather like a small Gothic (p. 73), this moth has blacker forewings and much paler hindwings. It comes very freely to light on August and September nights and is sometimes found sitting on grass-stems during the day. The female, which is larger and has darker hindwings than the male, lays her dingy white eggs on mat and other grasses, where they overwinter. The larvae hatch in the spring and feed at night until July, hiding by day in the soil, where they eventually pupate. Feathered Gothics are generally common in open country in England, Wales, and Ireland, but local in Scotland.

3 Antler Moth (*Cerapteryx graminis*). The grass-feeding larvae of this small moth sometimes appear in plague proportions in the high moorlands of northern Britain and eat acres of grassland. In normal years the moth is generally common in open grassy places almost everywhere in the British Isles. Both sexes have the characteristic white, antler-like marks on the forewings, but the female is distinctly the larger. They fly in July and August, frequently by day though more by night, when they are greatly attracted by light. The larvae hatch in early spring from batches of pale, shiny eggs, and feed on various grasses from March until June, when they pupate in the soil.

4 Common Wainscot (*Leucania pallens*). A very common moth of rough grassy places almost everywhere in the British Isles. The plain straw-coloured forewings render it inconspicuous when it rests by day among dead grasses or reeds. The female, which has less white hindwings, deposits her shiny whitish eggs in rows on various grasses and rushes, and the larvae feed on these from July through the winter until the spring, and then pupate in flimsy cocoons amongst the foodplants. The moths emerge in June and July. Some larvae feed up quickly and produce a second generation of moths in August and September, and these are generally smaller than the first generation.

5 Smoky Wainscot (*Leucania impura*). These are rather like the Common Wainscots, but with a black streak at the base of each forewing and smoky-grey hindwings. They usually fly in July and August in open country, especially in marshy localities, except in the extreme north. The large batches of shiny white eggs are deposited on various grasses on which the larvae feed from August, through the winter, and until May, when they change into chestnut-brown pupae in the soil.

6 The Clay (*Leucania lythargyria*). These are common moths of rough grassy places and marshes. Their forewings are a rather variable clay colour, ranging from light pinkish-brown to reddish-brown, with a conspicuous white dot and a cross row of small black dots on each. The moths fly at night in July and August and are much attracted to light. The larvae, which hatch from large batches of pale, shiny eggs, hide by day and feed at night on grasses and other low plants from August until they hibernate for the winter; they feed again in the spring and in May change to bright red-brown pupae in underground chambers.

7 The Shark (*Cucullia umbratica*). A moth with rather pointed, grey forewings; the hindwings of the male are white, but those of the female are clouded with dark scales. The pale, shiny eggs are deposited on sow-thistles and lettuce, on which the larvae feed at night from July to early September, when they pupate. The chestnut-brown pupae lie in the soil in earthen cocoons through the winter, and the moths emerge the following June or July and fly at night, visiting flowers. The moths usually rest by day on tree-trunks, walls, or in dark places. They are quite plentiful almost everywhere in the British Isles, especially in the south.

8 The Mullein (*Cucullia verbasci*). This is a browner moth than the Shark, with rather pointed wings and wing borders not unlike the edge of a postage stamp. The female, which has darker hindwings than the male, lays her dark-coloured eggs on mulleins, figworts, and occasionally on buddleia, on which the strikingly-marked larvae feed fully exposed from May to July. They are often particularly heavily parasitized by ichneumons (*see* p. 146). The greenish-brown pupae overwinter inside strong cocoons of silk interwoven with fragments of earth, and lying in the soil. The moths fly at night in April and May and sometimes come to light. They are difficult to see when at rest by day, owing to their effective camouflage. They are widespread and generally plentiful in England and Wales, especially the south, but are not found in Scotland nor Ireland, except for Co. Cork.

LIFE SIZE

1 Pine Beauty
2 Feathered Gothic
3 Antler Moth
4 Common Wainscot
5 Smoky Wainscot
6 The Clay
7 The Shark
8 The Mullein

For a general note on Noctuidae see p. 70.

1 The Sword-grass (*Xylena exsoleta*). Though more or less plentiful throughout the British Isles, the Sword-grass tends to be commonest in northern England and Scotland. There is a similar species, the Red Sword-grass (*X. vetusta*), which is commonest in Scotland and Ireland and rarest in England and Wales, and has a reddish tinge. The moths fly at night in the autumn and again, after hibernation, from March to May, when the female lays the batches of darkish eggs upon many kinds of plants and bushes. The larvae, which are large when full-grown, feed by day and night from April till July, and then pupate on the ground. The pupae are shiny chestnut brown.

2 Early Grey (*Xylocampa areola*). A predominantly grey moth, with curiously-shaped whitish marks on the forewings. It flies at night in March and April, visiting sallow catkins, and rests by day, perfectly camouflaged, on tree-trunks, walls, and fences. It is fairly common in wooded districts throughout the British Isles, especially in southern England. The larvae, which hatch from batches of dark, mottled eggs, feed at night upon honeysuckle from May until August, when they change into brown pupae in the soil, remaining there throughout the winter.

3 Merveille du Jour (*Griposia aprilina*). The green forewings of this lovely moth effectively camouflage it when it rests by day on lichen-covered trees. It flies by night from August to October in oak woods almost everywhere in the British Isles, though it is uncommon in Ireland. The female lays her dark-striped, pale eggs in batches on oak trees, where they overwinter. In the spring the larvae hatch and feed on the oak leaves until June, when they change into chestnut-brown pupae in subterranean cells at the base of the trees.

4 Brindled Green (*Dryobotodes eremita*). The usually greenish forewings of this moth are variegated with brown, black, and white. It is on the wing at night in September and October, and is generally common where oaks grow, except in the far north. In Ireland it is rare and restricted to the north and north-east. The females lay batches of light-coloured eggs, which over-winter, and the larvae feed from March to June on oak — at first on the buds, later on the leaves. In June they pupate in the soil at the base of oak trees.

5 The Satellite (*Eupsilia transversa*). A fairly common moth in wooded districts almost everywhere in the British Isles. There is a yellow, or sometimes white or orange, spot on each forewing, and the forewings are usually some shade of reddish or chestnut brown. Satellites fly at night in the autumn and again, after hibernation, in the spring, when they come to sallow catkins. The dark-coloured eggs are laid in April, mainly on the twigs of oak, elm, and beech, and the larvae feed on the leaves from April to June. They also eat other caterpillars. They form day shelters by spinning the leaves together with silk. They pupate in the soil beneath the trees.

6 Lunar Underwing (*Omphaloscelis lunosa*). The forewings vary from light yellowish-brown to blackish, normally with conspicuous lines of black dots near the outer edges; and the hindwings carry distinctive blackish marks. The moths are widespread and locally common throughout Britain, except in northern Scotland; they are scarcest in the east and most abundant in the south and west and all over Ireland. They fly at night in the autumn and come freely to light and ivy blossom. The batches of whitish eggs are deposited on grasses, such as annual meadow grass, and the larvae feed through the winter from September to May, when they change into brown pupae in the soil.

7 Red-line Quaker (*Agrochola lota*). All four wings of this moth are a dingy blackish-grey colour, and there is a distinctive red line near the outer margin of each forewing. It flies at night in open country from September to early November and likes to feed at the ivy blossom. Except in the extreme north, it is more or less common, especially in the south. The batches of dark, shiny eggs laid on willows and sallows overwinter and hatch in the spring, when the larvae feed until June on the leaves at night and hide by day in leafy shelters which they construct themselves. The brown pupae lie either between leaves drawn together with silk or on the ground until the moths emerge in the autumn.

8 The Brick (*Agrochola circellaris*). The light, tawny or yellowish-brown forewings are crossed with darker, wavy lines, and there is usually a dark spot on each, which helps in identification. The hindwings are greyish. Brick Moths fly at night from August to October, and are common in wooded districts almost everywhere in the British Isles. The yellowish eggs, which turn purple later, do not hatch until April, and then the larvae feed on the foliage, flowers, and seeds of wych elm and ash until June, when they pupate.

LIFE SIZE

1 THE SWORD-GRASS
3 MERVEILLE DU JOUR
5 THE SATELLITE
7 RED-LINE QUAKER

2 EARLY GREY
4 BRINDLED GREEN
6 LUNAR UNDERWING
8 THE BRICK

A general note on Noctuidae is found on p. 70.

1 Flounced Chestnut (*Anchoscelis helvola*). Although it varies a little, this moth is not unlike the Brick (p. 79), but has a dark transverse band near the outer margin of each reddish-brown forewing. The moths fly on autumn nights over moorlands in the north, where they are locally common, and also in wooded districts in southern Britain; but they are scarce in Ireland. The larvae, which hatch from overwintering batches of pinkish-brown eggs, feed in the spring chiefly on oak, elm, hawthorn, sallow and, in the north, bilberry and heather. The brown pupae lie in the soil.

2 Pink-barred Sallow (*Citria lutea*). This species is rather like the Sallow but has more heavily marked forewings. It flies on September and October nights and is generally common where sallows grow throughout the British Isles, apart from the extreme north of Scotland. The yellowish eggs change to purple, and are deposited on sallows, where they overwinter. The larvae hatch in March, feed on the sallow catkins until they fall, and then turn to the leaves of various low-growing plants. In June they change into brown pupae in the soil.

3 The Sallow (*Cirrhia icteritia*). This is less heavily marked than the Pink-barred Sallow and the ground colour of the forewings is lemon yellow; but its life history is similar, except that the eggs are dingy brown in colour, and the young larvae feed on aspen as well as sallow catkins. It is more or less common where sallows or aspens grow all over the British Isles, except in the extreme north.

4 The Chestnut (*Conistra vaccinii*). A small moth with chestnut-brown forewings, each marked with a blackish spot near the centre. It varies, however, considerably. The moths fly at night from September to December and again, after hibernation, in the spring, when they pair. They may be seen at ivy blossoms in the autumn and sallow catkins in the spring, and are usually very common in wooded districts throughout the British Isles. The larvae hatch from batches of shiny, pale eggs with dark markings, laid in bark crevices, and then feed in May and June upon oak, elm, and other trees, and also on lower-growing plants. The brown pupae lie in the soil at the base of the foodplant.

5 Marbled Beauty (*Cryphia perla*). This and the following species on this page belong to a group of the Noctuidae in which the larvae are all hairy and the majority of the moths some shade of grey, attractively marked with cryptic patterns. The Marbled Beauty has white forewings beautifully mottled with grey, which make it difficult to see on old stone walls, where it often rests in daytime. The female, which is larger than the male, lays her batches of shiny white eggs, banded with grey, usually on the lichen, *Lecidea confluens*, which grows on old walls. The larvae feed on the lichen from August until they hibernate for the winter in silken shelters spun up in crevices. In spring they reappear, feed until May, and then spin silken cocoons in other crevices and change into reddish-brown pupae. The moths appear in July and August and fly at night more or less commonly throughout the British Isles, except in northern Scotland.

6 Sycamore Moth (*Apatele aceris*). This and the Poplar Grey are much alike, though the Sycamore has lighter, mottled-grey forewings. The female has a stouter body than the male. They fly at night in June and July and are generally common in southern and eastern England, especially round London, but are rather uncommon in the Midlands, and are thought not to occur in Ireland. The attractive-looking larvae, which hatch from chequered blackish and white eggs, feed chiefly on sycamore and field maple leaves from July to September. The brown pupae pass the winter inside silken cocoons spun in crevices or cracks in the trunk of a tree.

7 Poplar Grey (*Apatele megacephala*). A darker-grey moth than the Sycamore with a noticeable white, black-centred spot near the middle of each forewing. It flies at night from May to August wherever poplars grow, and rests by day on tree-trunks. It is generally common in southern and south-eastern England, especially round London, and is found in many other areas in the British Isles. The speckled eggs are deposited on various kinds of poplars, on which the larvae feed from July to September. The brown pupae lie throughout the winter in silken cocoons constructed in crevices or under the bark of poplar trees.

8 Grey Dagger (*Apatele psi*). The black dagger-like markings on its plain grey forewings distinguish this moth, though it is almost impossible to tell from the less common Dark Dagger (*A. tridens*). The larvae of the Dark Dagger, however, are much less strikingly coloured. The moths fly on June and July nights and rest, well camouflaged, by day on tree-trunks or fences. They are generally common almost everywhere in the British Isles. The light-coloured eggs are laid on various trees and bushes, including lime, chestnut, hawthorn, blackthorn, and fruit trees, on which the larvae feed from July to October. The brown pupae pass the winter inside silken cocoons in crevices or under the bark of trees.

LIFE SIZE

1 FLOUNCED CHESTNUT   2 PINK-BARRED SALLOW
3 THE SALLOW   4 THE CHESTNUT
5 MARBLED BEAUTY   6 SYCAMORE MOTH
7 POPLAR GREY   8 GREY DAGGER

For a general note on Noctuidae see p. 70.

1 **Copper Underwing** (*Amphipyra pyramidea*). All the species on this and the next page normally have cryptically-coloured forewings, often with attractive patterns, which give good camouflage when they are resting by day (*see* p. 199). The females in most species have stouter bodies than the males. Unlike those on p. 80 their larvae are not hairy.
Copper Underwings fly from July to October in woodlands, and are locally common in the south, but rather uncommon and more localized in northern England and Wales. In Ireland they are widespread and generally plentiful. The dark-coloured eggs are deposited on various foodplants, including oak, elm, birch, sallows, rose, and brambles. They do not hatch until the following spring, when the larvae feed until June. They then form cocoons of silk and soil debris on the ground, in which they change to chestnut-brown pupae. The curved stump at the tail end makes it possible to mistake these larvae for small Hawk-moth larvae.

2 **The Mouse** (*Amphipyra tragopoginis*). When disturbed, this moth runs away in a mouse-like manner and is also much the same shade of brown as a House Mouse, It normally has two or three small black spots on each forewing, and the female is generally a little larger and stouter-bodied than the male. She lays her batches of dark-coloured eggs on many kinds of plants, including hawthorn and sallow, where they overwinter, and on the foliage of which the larvae feed from April to June. The brown pupae lie in the soil until the moths emerge in July or August. They are generally common almost everywhere in the British Isles.

3 **Old Lady Moth** (*Mormo maura*). A large moth often found by day hiding in the porch of a house or behind curtains, even in cities such as London; it is on the wing at dusk in July and August, when, like many big-bodied moths, it is preyed on by bats. It is especially common in southern England and fairly plentiful everywhere except in the extreme north. The female lays her shiny, white eggs upon low-growing plants, such as docks, on which the larvae feed in the autumn until they hibernate for the winter. In the spring they resume feeding on the young foliage of birch, hawthorn, sallow, and fruit trees until ready to pupate underground in May.

4 **Light Arches** (*Apamea lithoxylaea*). It is easy to confuse this moth with the less common Reddish Light Arches (*A. sublustris*) which has shorter, darker forewings with a reddish tinge. It flies in June and July and is generally common almost everywhere in the British Isles, except the extreme north. The batches of shiny white eggs are deposited on the seedheads of grasses such as annual meadow grass. The larvae feed from October until the following May on the grass stems, just above the roots. The brown pupae lie beneath the ground.

5 **Dark Arches** (*Apamea monoglypha*). Three examples of this very varied moth are shown here, of which the black form is becoming increasingly common. This may be an example of industrial melanism (*see* p. 199), an adaptation to an environment which is polluted by industrial grime. The darker forms, having more effective camouflage, survive better. Dark Arches are abundant almost everywhere in the British Isles and are easily attracted to light. They fly from June to August and sometimes again, as a second generation, in the autumn. The larvae, which hatch from eggs laid in small batches on various grasses, feed from July onwards. Some feed up quickly and produce autumn moths, but the majority overwinter and pupate underground the following June.

6 **Cloud-bordered Brindle** (*Apamea crenata*). This is another variable moth, rather like a smaller edition of the Light Arches, but the buffish-white forewings are more heavily clouded with brown. The forewings of some forms vary from creamy white to dark brown, and a blackish variety is spreading in some districts. The moths fly in plenty from May to July almost everywhere in the British Isles. The whitish eggs are laid in batches on the seedheads of various grasses on which the larvae feed from July until the following May. The brown pupae lie in the soil within flimsy earthen cocoons.

7 **Common Rustic** (*Apamea secalis*). Another very variable species, of which one of the commonest varieties is illustrated. It usually has a distinctive little white ring or spot near the centre of each forewing. It flies in July and August abundantly throughout the British Isles. The larvae hatch from rows of shiny white eggs deposited on the blades of various grasses, such as cock's-foot, and feed on the stems from August until the following spring, when they pupate in the soil. They also feed on wheat, sometimes to such an extent that they become a pest.

LIFE SIZE

1 Copper Underwing   2 The Mouse
4 Light Arches   3 Old Lady Moth
6 Cloud-bordered Brindle   5 Dark Arches
7 Common Rustic

For a general note on Noctuidae see p. 70.

1 Marbled Minor (*Procus strigilis*). A usually common little moth which varies a good deal in its markings, some forms being blackish or completely black. It is difficult to distinguish from two other Minors, except under the microscope. It flies at night in June and July almost everywhere in the British Isles, readily coming to light. During the day it rests well camouflaged on tree-trunks and fences. The female lays her shiny white eggs in dense rows on various grasses, especially cock's-foot, in the stems of which the larvae feed in the autumn and, after hibernation, in the spring until early May. The reddish-brown pupae lie in subterranean cells.

2 Flounced Rustic (*Luperina testacea*). The forewings vary somewhat in ground colour and markings, but the dark horizontal mark connecting the forewing's cross-lines and the white hindwings serve to distinguish it. It is usually common almost everywhere in open grassy places, particularly near the coast, and flies at night in August and September, when it is readily attracted to light. The female deposits her rows of shiny white eggs on various grasses, on the roots of which the larvae feed during the autumn and again, after hibernation, in the spring. They are sometimes a pest of wheat. The light-brown pupae lie beneath the grass roots.

3 The Angle Shades (*Phlogophora meticulosa*). When resting by day, this lovely moth, with its delicate colouring and angle-shaped markings on the forewings, strongly resembles a shrivelling leaf. Angle Shades fly at night, are generally common all over the British Isles, and have been seen in most months of the year. They are most numerous from May to July and again from mid-August to October, so presumably there are two generations each year, and there are also many immigrants, especially in the autumn. The larvae can be either green or brown; they occur at almost any time of year, hatching from shiny, mottled eggs, but are most common in spring and late summer, and they feed on many kinds of wild plants, as well as many garden ones. They most often hibernate as larvae. The chestnut-brown pupae lie in flimsy silken cocoons on the ground.

4 Straw Underwing (*Thalpophila matura*). This species is recognized by its dark-bordered, yellowish hindwings. It is more or less common in rough, grassy places over most of the British Isles except the extreme north, and flies on July and August nights when it is easily attracted to light. The larvae, which hatch from darkish ribbed eggs, feed on various grasses in dry situations from late August until they hibernate; then they feed again until they pupate in the ground in May.

5 Pale Mottled Willow (*Caradrina clavipalpis*). This small, pale-brown moth has white hindwings and distinctive black dots on the front edges of the forewings. The female is slightly larger than the male. These moths have been recorded in every month from February to November, though most commonly from May to September, and there are at least two generations each year. They are generally abundant throughout the British Isles. The larvae, which hatch from pale-coloured eggs, generally feed from September until May on chickweed, various grasses, peas, and plantain seeds, and are often a pest in grain stacks. The brown pupae lie on or in the soil in cocoons spun together from earth and debris.

6 Rosy Rustic (*Gortyna micacea*). A generally rosy-brown moth with characteristic markings, which is on the wing at night in the late summer and autumn. It is usually common in damp, open country throughout the British Isles, and is attracted to light. The shiny, dark eggs overwinter, and the larvae feed from May to August in the stems and roots of plantains, docks, potatoes, tomatoes, and other plants. The brown pupae lie in underground cells.

7 The Dun-bar (*Cosmia trapezina*). The larvae of this extremely variable moth are notorious cannibals, attacking and eating other caterpillars; otherwise they feed on the foliage of various trees, such as oak, elm, birch, hazel, and sallow, from April until June, when they pupate on the ground within flimsy silken cocoons. The moths emerge from glossy brown pupae in July and August and fly at night, more or less commonly, in wooded districts over most of the British Isles, except the far north. The batches of dark eggs overwinter and hatch in the spring.

8 Bulrush Wainscot (*Nonagria typhae*). The pale, brownish-white forewings, dotted and streaked with black, make this moth inconspicuous when resting by day among the reed-mace and reeds in its watery habitat. It is larger than most Wainscot moths (*see* p. 77). The female lays her eggs in the stems of reed-mace in slits which she makes with her specially adapted ovipositor, and they hatch the following spring. The larvae feed on the pith inside the stems until they pupate in July or August in partitions in the stems which they seal off. The dark-brown pupae hang head downwards. The moths emerge in August and September and fly at night wherever reed-mace grows profusely in England and Wales as far north as the Lake District, and in Ireland, and are plentiful in most districts.

LIFE SIZE

1 MARBLED MINOR   2 FLOUNCED RUSTIC
3 THE ANGLE SHADES   4 STRAW UNDERWING
5 PALE MOTTLED WILLOW   6 ROSY RUSTIC
7 THE DUN-BAR   8 BULRUSH WAINSCOT

For a general note on Noctuidae see p. 70.

1 Cream-bordered Green Pea (*Earias clorana*). This moth can be distinguished from the Green Oak-roller Moth (p. 121) by the white instead of grey hindwings, the creamy-white front edges of the forewings, and the white head. It flies on May and June nights in marshy places where osiers and willows grow, and is generally common in the south and east of England, but extremely rare in Ireland. The larvae feed from late June to August among the terminal leaves of osiers and willows, spinning them together with silk. The dark-brown pupae overwinter in strong, light-brown, boat-shaped cocoons attached to a twig.

2 Green Silver Lines (*Bena fagana*). A lovely moth which can be distinguished from the rare Scarce Silver Lines (*Pseudoips prasinana*) by the three instead of two silver-white cross-lines on the forewings. The female is larger than the male and has pure white instead of yellowish hindwings; she deposits her eggs upon oak, beech, birch, or hazel, and the larvae feed on the foliage from late July to September. When full-fed, the larvae construct strong, pinkish-brown, boat-shaped cocoons in a variety of situations, such as under a leaf, on bark, or amongst the leaf litter on the ground, within which the purple-tinged brown pupae lie through the winter. The moths emerge the following June or July, and fly by night in woodlands. They are locally plentiful in many parts of the British Isles, especially England.

3 Red Underwing (*Catocala nupta*). A large, handsome moth which, like the next two species, has attractively coloured hindwings. This is one of the largest British moths and rests by day usually high up on tree trunks, or on posts and old walls, where its cryptically-coloured forewings make it not easy to see. When disturbed, it flies off rapidly with a sudden display of brilliant red hindwings, which the enemy pursues. Then, in the same way as the Lesser Yellow Underwing does (p. 72), it settles again, and the red wings disappear. It normally flies at night, in August and September, and frequently comes to light. It is generally plentiful wherever poplars and willows abound in southern and eastern England, and occasionally occurs further north, though not in Scotland or Ireland.
The female lays her purple-brown eggs singly in fissures in the bark of poplar and willow trees, on the leaves of which the larvae feed at night after hatching from the eggs in April. They rest by day on the bark, which they match extremely well. They pupate in July, constructing cocoons of rough silk among the leaves. The brown pupae are coated with a bluish-white powder.

4 Mother Shipton (*Euclidimera mi*). The markings on the forewings of this moth are said to resemble the profile of a wizened old woman — hence its name. The moths fly by day in May and June, frequenting open grassy places and fields rich in flowers all over the British Isles, except the north of Scotland. The dull-green eggs are deposited in clusters on various clovers, upon which the larvae feed from July to September. They overwinter as chestnut-brown pupae, powdered with white, hidden inside brownish-grey, silk cocoons attached to the grass.

5 Burnet Companion (*Ectypa glyphica*). A moth with a butterfly-like flight, which flies by day in much the same habitats as the Mother Shipton and, as its name indicates, the Burnet moths (p. 112), and readily visits flowers. It flies in May and June and is locally common in many parts of the British Isles, though commonest in southern England. The larvae, which hatch from green eggs laid in batches, feed at night on clovers from June to August. The purplish-brown pupae overwinter on the ground in strong cocoons of silk and pieces of plant.

6 Golden Plusia (*Polychrisia moneta*). This, the next species, and the first three species on p. 88 belong to a group which often have beautiful metallic and prominent tufts or ruffs of scales on their thoraxes. The Golden Plusia was unknown in the British Isles before 1890, but since then it has extended its range all over England, Wales, and southern Scotland, and is now common in the south. It reached Ireland in 1939 and is now locally plentiful in Co. Dublin. It flies on June and July nights, and sometimes a second generation flies in August and September. It is most likely to be seen in parks and gardens where its foodplants, larkspur and monkshood, grow. The larvae, which hatch from shiny white eggs laid on the foodplants, hibernate when still young and feed the following May and June. In years when second generations occur, larvae may also be found in July and August. The light-brown pupae are contained within yellow silken cocoons attached to the undersides of leaves.

7 Burnished Brass (*Plusia chrysitis*). There are two metallic brass-golden bands on each forewing, from which the moth is named. It flies at night, and sometimes during the day, between June and September, visiting flowers for their nectar; it is common in open country, especially where nettles grow, almost everywhere in the British Isles, except the extreme north. The larvae, which hatch from yellowish-white eggs, feed chiefly on stinging nettle and hibernate when still small. They reappear in spring and pupate in May within silken, white cocoons attached to the foodplant. The pupae are black. In some years a second generation of larvae in late summer produce moths in the autumn.

LIFE SIZE

1 Cream-bordered Green Pea   2 Green Silver Lines

3 Red Underwing

4 Mother Shipton   5 Burnet Companion

6 Golden Plusia   7 Burnished Brass

NOCTUAS (*Noctuidae*) AND LACKEY (*Lasiocampidae*)

For a general note on Noctuidae see p. 70.

1 Silver Y (*Plusia gamma*). The conspicuous silvery-white mark on each forewing is the reason for both the common and scientific names of this moth, for the marks resemble both the letter 'Y' and the Greek letter *gamma*. The moths are not easy to see when sitting in the vegetation, but are readily attracted to artificial light at night and may also be seen on the wing by day. They visit flowers, especially at dusk. Silver Ys do not normally survive the British winter, but each spring immigrants fly over from southern Europe and breed in Britain, giving rise to a large autumn generation of moths: in particularly good years they appear in enormous numbers. They are usually common in almost every part of the British Isles from May to October. The larvae, which hatch from pale, rather shiny eggs, vary in colour from green to dark olive and feed on a great variety of cultivated and wild plants between June and September. When ready to pupate, they spin white, silken cocoons attached to the food-plant, often under a leaf; the moths emerge about three weeks later from the blackish pupae.

2 The Spectacle (*Unca triplasia*). The two black-ringed, grey markings on the front of the thorax, looking very like a pair of spectacles, account for this moth's name. In the closely related, but darker, Dark Spectacle (*U. trigemina*) the spectacle marks are reddish-brown. The Spectacle Moth flies at night, visiting flowers, in May and June, and a second generation flies in August; they are locally common throughout the British Isles, except the Shetlands. The larvae, which hatch from pale-coloured eggs, feed on stinging nettle at night in July and again in September. The pupae lie inside whitish cocoons spun under a nettle leaf, and overwinter in this stage.

3 Figure of Eight Moth (*Episema caeruleocephala*). One of the white marks on each forewing of this moth normally resembles a figure 8. The female, which lacks the male's feathery antennae, deposits her small batches of blackish-grey eggs on the side shoots of blackthorn, hawthorn, crab-apple, and fruit trees, where they remain through the winter. In April the larvae emerge and feed on the foliage until they pupate in June. They construct tough cocoons composed of silk and pieces of the foodplant under twigs or on the bark. The moths emerge from the light purple-brown pupae in the autumn and fly at night. They are found throughout England, Wales, southern Scotland, and northern Ireland, but are rather uncommon except in southern England.

4 The Herald (*Scoliopteryx libatrix*). A handsome moth resembling a shrivelled leaf when resting during the daytime. It flies at night from August to October, and again, after hibernation, in the spring. It visits ivy blossom in the autumn and sallow catkins in the spring, and often hibernates in buildings or hollow trees. The female, which lacks the male's feathery antennae, lays her pale, shiny eggs in the spring upon sallows, willows, and poplars, and the larvae feed on the foliage from May to August. The larvae spin white cocoons between the leaves of the foodplant and change to black pupae within. Herald moths are generally plentiful almost everywhere in the British Isles, especially in the south.

5 Small Purple-barred (*Phytometra viridaria*). The shade of brown on the wings of this little moth varies somewhat, and the characteristic rose-red bands are sometimes purplish. It is found almost everywhere in the British Isles where its foodplant, common milkwort, grows, and is usually plentiful. It flies in May and June, and the larvae, which hatch from pink-mottled white eggs, feed on the milkwort from July until September, when they spin grey silk cocoons amongst parts of the foodplant. They pass the winter as pupae.

6 The Snout (*Hypena proboscidalis*). Moths of this group have long palps (sense organs) which stick out in front of the head like a snout. Snouts are generally common throughout the British Isles and fly at night in June and July, and sometimes again in the autumn as a partial second generation. They are often disturbed during the day from nettle beds. The larvae feed on stinging nettles normally from August until they go into hibernation for the winter; in the spring they reappear and pupate in June, probably among the leaves of the foodplant.

7 The Lackey (*Malacosoma neustria*). This species belongs to the family Lasiocampidae, about which there is a general note on p. 90. Lackeys fly at night in July and August and are found in open country throughout England and Wales as far north as Lancashire and Yorkshire, but they are only really common in the south. They are rare in northern Ireland, but locally plentiful in the south. They vary in colour, the male being usually yellowish or reddish-brown, and the female, which is larger and stouter-bodied, is pale to reddish brown. The shiny brown eggs are laid in a batch encircling a twig of the foodplant and do not hatch until the following April. The brightly-coloured larvae live gregariously in conspicuous webs until they are nearly full-grown, and feed on a variety of trees, especially blackthorn, hawthorn, oak, and hazel. Their bright colours are a warning to predators that they are distasteful. In June, each larva spins a double silken cocoon among the leaves, in which it changes into a blackish pupa, sprinkled with yellow powder.

LIFE SIZE

1 Silver Y   2 The Spectacle
3 Figure of Eight Moth   4 The Herald
5 Small Purple-barred
6 The Snout   7 The Lackey

The Lasiocampidae is a family of generally medium or large sized brown moths with stout bodies and either no or only rudimentary tongues. Their larvae are all very hairy.

1 **Oak Eggar** (*Lasiocampa quercus*). There are two sub-species of this Eggar: a northern and darker one, *callunae*, often called the Northern Eggar, of which the female is shown here (top right), and a southern one, *quercus*, of which both sexes are shown. The Northern Eggar could be confused with the much more local Grass Eggar (*L. trifolii*), which, however, is smaller and has no yellow band on the hindwings. The male *callunae* also differs from the male *quercus* in having yellowish patches at the base of the forewings. In both subspecies the males are much smaller and darker than the females, and have plumed antennae.

Oak Eggars are found throughout the British Isles and are locally common. The northern subspecies inhabits the heather moors of Ireland and northern Britain as far south as Lancashire, Yorkshire, and north Wales, and also occurs on Dartmoor and Exmoor and, occasionally, elsewhere in England; and the southern one frequents heathlands, wood edges, hedgerows, and coastal dunes anywhere in the midlands and south.

The males fly by day in a fast, zigzag manner, seeking out the unmated females by their scent. After mating, the females fly about, dropping at random in the vegetation below their greyish eggs, mottled with light brown. The larvae of *quercus* hatch in August or September, soon go into hibernation, and next spring feed on a variety of plants, including bramble, broom, sallow, and hawthorn, until June or July, when they normally pupate, the moths emerging in July and August. Those of *callunae* feed on heather as well as these other plants, and do not normally pupate until the autumn; they pass the second winter in the pupal state and emerge in the following May or June. The pupae of both are purplish-brown, contained within strong, yellowish, egg-shaped cocoons which lie in strands of silk on the ground amongst the vegetation.

2 **Fox Moth** (*Macrothylacia rubi*). A more or less common species, especially as larvae, on commons, heathlands, and moors almost everywhere in the British Isles. The female is greyer than the male, although his normal shade of reddish brown may vary. The males fly on sunny days in May and June in the same wild and rapid manner as the male Oak Eggars, and seek out the unmated females by their scent. The females fly after dark and lay their batches of grey and brown eggs on the stems of heather, heaths, bilberry, and bramble, on which the larvae feed. They hatch during July and hibernate, fully-fed, during the winter. The first warm sunshine in early spring brings them out again, but they do not feed. They spin elongated, brownish cocoons mixed with their own hairs on or near the ground, in which they pupate in late March or April, the moths emerging a month or two later. The larvae should be handled carefully as their easily detachable hairs may cause a skin rash.

3 **The Drinker** (*Philudoria potatoria*). The larva of this moth has a habit of drinking dew and raindrops, hence its name. The male moth is usually slightly smaller and darker than the female, and also has feather-like antennae. Both sexes have a distinctive oblique black line and two small silver-white marks on each fore-wing. When at rest during the daytime, they could easily be passed over for a bunch of dead leaves. They fly at night in July and are attracted to artificial light. They are found in open country almost everywhere in the British Isles, though most commonly in damp or marshy localities where reeds and lank grasses grow. The shiny white eggs, marked with grey or grey-green, are deposited in groups upon the stems of common reed and various grasses and sedges, on which the hairy larvae feed on hatching in August. They hibernate from October until the following April, when they reappear and feed again until they pupate in June. The brown pupae are enclosed within long, papery, ochre-coloured or light-brown cocoons attached to reed or grass stems.

4 **The Lappet** (*Gastropacha quercifolia*). Its English name refers to the fleshy lappets on each side of the body of the larva. These reduce the shadow of the body, making the larva less conspicuous. Like the Drinker, the moths resemble dead bunches of leaves when resting during the day. The female is much larger and darker than the male and has less feathery antennae. She deposits her eggs, which are whitish, spotted and banded with grey, in pairs or small groups on the twigs or leaves of blackthorn, hawthorn, fruit trees, sallows, and, occasionally, other trees on which the larvae feed from August until they hibernate for the winter at the base of the foodplant. Next April they resume feeding and become very large by the time they pupate in late May or June. The larva constructs a long, grey cocoon of silk mixed with its own hairs, attaches it to the food-plant, and changes within to a blackish-brown pupa which is covered with white powder. The moths emerge about a month later and fly rapidly at night in June, July, and August along hedgerows and wood borders in fairly open country. It is restricted to England, but is widespread there, though much less plentiful in the north.

LIFE SIZE

1 OAK EGGAR   2 FOX MOTH
3 THE DRINKER   4 THE LAPPET

The Arctiidae are a family of moths which vary a good deal in size and appearance. There are two sub-families, the Footmen and the Tigers and Ermines, of which three are shown on p. 95. They all have hairy larvae. The Footman moths have well-developed tongues and visit flowers, which the Tigers do not, as their tongues are poorly developed.

1 **Red-necked Footman** (*Atolmis rubricollis*). The name 'Footman' is used because some of these moths wrap their long, narrow wings tightly around themselves when resting during the daytime, so suggesting the stiff appearance of a footman in livery.
The Red-necked Footman is locally plentiful in woodlands in the southern half of England and in Ireland, though rare elsewhere. It flies in June and July at night, and sometimes on sunny days, when it feeds freely at flowers. The female deposits her eggs in small batches on coniferous trees, oaks, and beeches wherever the green alga *Pleurococcus naegelii* grows. The hairy larvae feed on this alga from July until October, hibernate for the winter, and then resume feeding until May, when they spin silken cocoons mixed with their own hairs, amongst lichen or in crevices in the bark, and change within to dark chestnut-brown pupae.

2 **Common Footman** (*Lithosia lurideola*). A similar species to the Scarce Footman (*L. complana*), but distinguished by the yellowish band along the leading edge of the forewings; in the Scarce Footman this extends right to the wing-tip. The female deposits her batches of shiny eggs on various algae and lichens growing on trees and rocks, and on which the larvae feed from August until the following June. In captivity the larvae will feed on the leaves of oak, sallow, and other trees. They spin silken cocoons amongst lichens or in bark crevices, and change within to chestnut-brown pupae. The moths, which are fairly common in the southern half of Britain, fly at night in July and early August. They are much more local and scarce further north, though locally plentiful in Ireland.

3 **White Ermine** (*Spilosoma lubricipeda*). A common moth of gardens and open country all over the British Isles, flying by night from mid-May to July. The size and number of the black spots on the white wings vary a good deal. The female is stouter-bodied and has thinner antennae than the male. She deposits her shiny white eggs in large, orderly batches under the leaves of dandelion, dock, and other low plants, on which the very hairy larvae feed from July to September. They pass the winter amongst the ground litter as dark-brown pupae enclosed in cocoons of silk mixed with their own hair.

4 **Buff Ermine** (*Spilosoma lutea*). A common moth throughout the British Isles, which varies a good deal in the shade of buff and the intensity of the black markings. The females are generally paler than the males and have stouter bodies and thinner antennae.

Their life history is similar to the White Ermine, but the eggs are greenish and the pupae chestnut. The moths fly at night in June and July. The larvae feed on many plants, particularly docks.

5 **Muslin Moth** (*Cycnia mendica*). The sexes of the typical race are quite different, the male being sooty or blackish brown and the female white with black spots. In the Irish race (*rustica*) both males and females are white or sometimes cream or ochre coloured, though the males have thicker antennae. They are locally plentiful throughout the British Isles, except in the extreme north, and fly at night in May and June, coming freely to light. The larvae, which hatch from batches of shiny white eggs, feed on dandelion, docks, and many other plants from June to August, and pass the winter on the ground as blackish-brown pupae within compact silken cocoons mixed with their own hairs.

6 **Ruby Tiger** (*Phragmatobia fuliginosa*). A common moth almost everywhere in the British Isles, which flies by night, and sometimes by day, from April to June, and again as a second generation from July to early September. It frequents open woods, heaths, moors, marshy meadows, and other grassy places, and the stouter females deposit their large batches of pearly-white eggs on heather, docks, dandelions, and other plants. The first-generation larvae feed in May and June, and those of the second generation from July to October and again, after hibernation, in the early spring, when they pupate. The black pupae, which have yellowish rings on the hind segments, lie within brown, silken, oval cocoons spun amongst the ground vegetation. In northern Britain and in Ireland the moths are darker than in the south and much less red.

7 **Wood Tiger** (*Parasemia plantaginis*). This moth varies a good deal; in mountainous areas in the north the male may have an entirely white ground colour (var. *hospita*). The males have more feathery antennae than the females. The Wood Tiger is fairly common in woodland clearings and on heaths, moors, and chalk and limestone hills throughout the British Isles, and is on the wing on sunny days and also at night from late May to July. The larvae hatch from the batches of shiny, yellowish eggs and feed on plantains, wild violets, dandelion, and other plants from July until they hibernate. They resume feeding in the spring, and pupate in May. The dark-brown pupae are contained within grey silken cocoons, mixed with larval hairs and spun on or near the ground.

LIFE SIZE

1 RED-NECKED FOOTMAN   2 COMMON FOOTMAN
3 WHITE ERMINE   4 BUFF ERMINE
5 MUSLIN MOTH
6 RUBY TIGER   7 WOOD TIGER

For a general note on Tiger Moths see p. 92, and for Geometer moths see p. 96.

1 Garden Tiger (*Arctia caja*). This moth has the bright colours which the layman usually associates with butterflies. These colours, and also a distinctive smell, warn predators, such as birds, that the moth is distasteful. When attacked, the Tiger extends its antennae and exposes its bright-coloured hindwings and the fringe of red hairs behind the head. Although typical specimens are unmistakable, Garden Tigers do vary a good deal. The female is larger and stouter-bodied than the male and has slimmer antennae. She deposits her batches of white or green eggs on the undersides of the leaves of the foodplant. The very hairy larvae, commonly called 'woolly bears', feed on the foliage of many kinds of plants, including docks, dandelion, and stinging nettle, from August until they hibernate while still small. They reappear in the spring and, when ready to pupate in June, construct loose, silken cocoons mixed with their own hairs amongst the ground vegetation. The moths emerge from the black pupae in July or August and fly at night, the males frequently coming to light. Garden Tigers are generally common in open country, especially gardens, throughout the British Isles.

2 Cream-spot Tiger (*Arctia villica*). A generally plentiful moth of hedgerows, woodland clearings, marshes, and such kind of open country in southern England and East Anglia, but not elsewhere. The sexes are alike, except that the female is larger than the male and possesses thinner antennae. The moths fly at night in late May and June and are attracted to light. The shiny pearl-white eggs are laid in neat batches on groundsel, ragwort, chickweed, dead nettle, dock, and other plants, on which the hairy larvae feed from July until the autumn when, still young, they hibernate for the winter. In the early spring they resume feeding and during May spin loose, grey, silken cocoons amongst the ground vegetation and pupate inside. The pupae are blackish.

3 Cinnabar (*Callimorpha jacobaeae*). This moth is an excellent example of a warningly coloured insect, both as an adult and as a larva. It is unpleasant to the taste, and birds, for example, soon learn to recognize it and leave it alone. The moths, which appear in May and June, often fly sluggishly by day, but more usually at dusk, when they are sometimes attracted to light. Except that the female has a stouter body, the sexes are alike. The conspicuous larvae differ from others of the family by being much less hairy. They hatch from large batches of shiny yellow eggs, feed gregariously

and openly in July and August on common ragwort, or sometimes on Oxford ragwort, groundsel, and coltsfoot. They often completely denude the plants of foliage and serve a useful purpose in controlling the spread of ragwort. In fact, attempts have been made deliberately to introduce them for this purpose, especially into New Zealand. The dark-brown pupae overwinter in thin silken cocoons which lie just in the soil or amongst the surface litter. The Cinnabar is generally common wherever it is found and inhabits open, rather dry country where ragwort flourishes. It occurs throughout the British Isles, except for the extreme north of Scotland, and is often numerous on waste ground in cities such as Bristol and London.

4 Orange Underwing (*Archiearis parthenias*). This Underwing should not be confused with the Underwings of the large Noctuidae family (*see* p. 72). This and the next species belong to a sub-family of Geometers (*see* p. 96). It can be mistaken for its close relative, the rarer Light Orange Underwing (*A. notha*), which is smaller and has plainer forewings. The males, which are generally rather darker than the females, fly in open birch woods or on heathland where there are birches, on sunny days in March and April, and can be seen around the tree-tops, where the females are sitting. They are widespread in England as far north as Co. Durham, but are common only in the south and east; they are very local and scarce in Scotland and Wales and unknown in Ireland. In April the young larvae feed on birch catkins, and later change to the leaves. In early June they descend the trees and pupate in flimsy silk cocoons, either on the ground or in bark crevices. They overwinter as pupae, which are dark chestnut brown.

5 March Moth (*Alsophila aescularia*). A drab little moth which is the only British representative of its sub-family. The winged males and wingless females may be found by day from February to April resting on tree-trunks and fences; the males fly at night and often come to light. The larvae hatch from dark, shiny eggs laid in large clusters on the twigs of oak, hawthorn, and various other trees and shrubs, and feed on the foliage from April to June. The brown pupae, enclosed in flimsy cocoons, lie in the soil beneath the foodplant throughout the winter. Except for the north of Scotland, the March Moth is generally common all over Britain, and is locally plentiful in Ireland, especially in the north.

LIFE SIZE

1 GARDEN TIGER
2 CREAM-SPOT TIGER   3 CINNABAR
4 ORANGE UNDERWING   5 MARCH MOTH

GEOMETERS (*Geometridae*)

The Geometers, a huge family with several sub-divisions, are mostly small or medium-sized, weak-flying, and inconspicuously-coloured moths. They generally fly at night, but also in the day-time if disturbed from their hiding places. Most rest with their wings spread flat, though some rest like butterflies, with wings raised and closed over the back. The sexes, unless otherwise stated, are alike, though the females have shorter, stouter bodies. The typical Geometer larvae possess only two pairs of claspers (hind legs) and arch or loop their bodies when travelling, so are called 'looper' caterpillars. They appear to be measuring the ground — hence the name 'Geometer', ground-measurer. They often stand erect on their claspers, looking like sticks.

1 **Grass Emerald** (*Pseudoterpna pruinata*). The members of this sub-family (Geometrinae) are usually some shade of green. Not long after emergence, the lovely blue-green colour of the Grass Emerald fades to a duller hue; consequently the moths are not easy to see when resting in June or July amongst the vegetation on their native heaths and moors, where they are common all over the British Isles, except the extreme north. The larvae, which hatch from batches or piles of creamy-white eggs, feed on gorse, broom, and needlewhin from August until they hibernate, and again in the spring until May when they pupate. The pupae lie in fragile silken cocoons on the ground amongst the leaf litter.

2 **Large Emerald** (*Geometra papilionaria*). This is the largest of the Emeralds, the female being a little the larger. She deposits her whitish eggs, which soon become greenish yellow, chiefly on birch, but also on hazel or beech. The larvae hatch in late summer, and in the autumn hibernate on a twig, near a bud, until the spring, at which stage they are mainly reddish-brown — the colour of the bud. They gradually become greener as the foliage grows, until by May they are completely green. So at all stages their camouflage is perfect. Cocoons containing rather large green pupae lie amongst the leaf litter on the ground.
The Large Emerald is common in woods and on heathland where birches are plentiful, all over the British Isles except for the far north. It flies in the latter part of the night in June and July, and also, if disturbed, during the day.

3 **Blood-vein** (*Calothysanis amata*). The name refers to the characteritsic reddish cross stripe. The moth is on the wing in June, and sometimes a second brood flies in August or September. The larvae feed upon various low plants, particularly docks, from July and again after hibernation until May, though some larvae feed up quickly, pupate, and produce moths in the autumn. Like others of this sub-family, the pupae lie within fragile cocoons amongst the ground litter. Blood-vein moths haunt hedgerows and places where their food-plants are plentiful throughout the British Isles, especially in southern England, but are rare in Scotland and Ireland.

4 **Maiden's Blush** (*Cosymbia punctaria*). A common moth in southern England, becoming scarce further north and very rare in Ireland. It flies in woods in May and June, and again, as a second generation, in August. The larvae feed on oak in June and July, and again in the autumn, and they overwinter in the pupal stage in flimsy cocoons among dead leaves on the ground.

5 **Cream Wave** (*Scopula lactata*). This moth flies in woods in May and June. It is rather rare in Scotland and only locally common in Ireland. There is only one generation, and the larvae, which hatch from reddish eggs, feed on low-growing plants such as bedstraw, woodruff, and dock from July until September. They hibernate, and in the spring pupate in the soil. The pupae are chestnut-brown.

6 **Riband Wave** (*Sterrha aversata*). Though usually greyish white, there is a brownish-yellow form with a wide, dark band crossing each wing. It flies in June and July, and there is sometimes a second brood in August and September; it is common throughout the British Isles, except the extreme north. The larvae feed on plants such as dandelion, dock, and bedstraw, and then, after hibernation, upon the young foliage of sallow, hawthorn, and other trees. They pupate in May in the soil. Some, however, may feed up quickly and pupate in the late summer, producing the autumn moths.

7 **Silver-ground Carpet** (*Xanthorhoë montanata*). A common moth of wayside and woodland throughout the British Isles. It flies in June and July, and the larvae feed at night on plants such as bedstraw, and on grasses. They pupate in the ground in April after hibernation. (See note on Carpets, p. 98.)

8 **Garden Carpet** (*Xanthorhoë fluctuata*). An abundant and familiar Carpet moth (*see* p. 98), even in town gardens, all over the British Isles from April to October, though commonest in late spring and late summer. There are at least two generations. The larvae feed at night from June to October on cabbage and other Cruciferous plants. The brown pupae lie in the soil within a silken cocoon.

LIFE SIZE

1 GRASS EMERALD   2 LARGE EMERALD
3 BLOOD-VEIN   4 MAIDEN'S BLUSH
5 CREAM WAVE   6 RIBAND WAVE
7 SILVER-GROUND CARPET   8 GARDEN CARPET

There is a general note on the Geometers on p. 96. The Carpet moths (pp. 46, 100, and 102) and the Pugs (p. 104) belong to this sub-family of small moths, which rest with their wings spread out, some species partially exposing the hindwings. The eggs are generally creamy-white and often shiny, and the larvae of most species tend to rest in a hooped position looking rather like a question mark. The thin, pointed pupae have projecting heads in which the eyes of the future moth can be seen.

1 **Red Twin-spot Carpet** (*Xanthorhoë spadicearia*). A pretty but variable moth of reddish-brown and grey colouring, with twin black spots on each forewing. It is common in most parts of England, Wales, and Ireland, scarcer further north, and not found in northern Scotland. It flies in May and June, and often again in the south as a second generation in July and August. The larvae feed in June and July, and also in the autumn, on various plants such as dandelion, ground ivy, knotgrass, and bedstraws. The smooth brown pupae lie in fragile silken cocoons at the base of the foodplant throughout the winter.

2 **The Mallow** (*Larentia clavaria*). This rather large Geometer, which is seen even as late as November, can be confused with the common Shaded Broad-bar (*Ortholitha chenopodiata*), but the latter does not fly so late. The Mallow flies quite rapidly at dusk wherever its foodplant, mallow, grows in the British Isles. It is commonest in southern England, scarce and local in Ireland, and not found in the extreme north. The larvae eat hollyhock as well as mallow. Mallows hatch in the spring, pupate underground in late June, emerge in the autumn, and overwinter as eggs.

3 **Green Carpet** (*Colostygia pectinataria*). A common moth almost everywhere in the British Isles, but the green forewings fade quite soon after emergence. It flies at dusk in June and sometimes again, as a second brood, in August and September, frequenting hedges and open country. The larvae feed chiefly upon bedstraws from late summer until May and those which feed up quickly produce a second brood; they overwinter as larvae.

4 **Twin-spot Carpet** (*Colostygia didymata*). A greyish or brownish little moth, with twin black spots on each forewing. It is common throughout the British Isles, especially in the moorland areas of the north and in Ireland. The female is slightly smaller and paler than the male. It flies in July and August and overwinters in the egg stage. The larvae feed at night in the spring on primrose, red campion, whortleberry, and other plants, and pupate amongst ground litter in June.

5 **Shoulder-stripe** (*Earophila badiata*). This moth pupates underground and emerges in March and April. It is common throughout the British Isles where wild roses grow, except for the far north. The larvae feed at night from May until July upon the leaves of dog and other wild roses.

6 **The Streamer** (*Anticlea derivata*). Another spring moth which is found where there are wild roses almost everywhere in the British Isles, though scarce in Ireland. The moths fly at dusk along the hedgerows in April and early May, and the eggs are deposited upon the buds or stems of wild roses or, sometimes, honeysuckle. The larvae feed in May and June and pupate in June or early July.

7 **Beautiful Carpet** (*Mesoleuca albicillata*). A common moth in June and July in open woodlands in southern England, and also found over most of England, Wales, southern Scotland, and Ireland. The larvae feed at night on bramble, raspberry, and wild strawberry from July to September or early October, and then change into dark chestnut-brown pupae just below ground or amongst the leaf litter on the surface.

8 **Rivulet** (*Perizoma affinitata*). A small moth, though larger and browner than the Small Rivulet (*P. alchemillata*), which flies at dusk in June and July and is common in many districts of England, Wales, and Scotland, except the far north; in Ireland it is found only locally in the north. The larvae feed from July to September in the seed-capsules of red campion, and also sometimes of white campion and ragged robin.

9 **Yellow Shell** (*Euphyia bilineata*). This moth is abundant in hedgerows and similar places throughout the British Isles. The blackish markings vary considerably in intensity, and dark blackish-brown and almost black forms are found in western Ireland; there is a small, dark race in the Hebrides and Shetlands. The moths fly after dark, and also frequently in the daytime, from June to August. The larvae, which hatch from tiny yellowish eggs deposited on the foliage, feed at night on grasses, chickweeds, docks, and other plants, and hide near the ground by day. They overwinter as larvae and pupate in May.

LIFE SIZE

1 RED TWIN-SPOT CARPET   2 THE MALLOW
3 GREEN CARPET   4 TWIN-SPOT CARPET
5 SHOULDER-STRIPE   6 THE STREAMER
7 BEAUTIFUL CARPET
8 RIVULET   9 YELLOW SHELL

There is a general note on Geometers on p. 96, and on the Carpet sub-family on p. 98.

1 Pretty Chalk Carpet (*Melanthia procellata*). A common moth on limestone soils in southern England, where its foodplant, traveller's joy, grows, but rarely found elsewhere. It can be confused with the Beautiful Carpet (p. 98) but is browner. It flies at night in July and early August, but also in the daytime if disturbed from its hiding places amongst or near traveller's joy. The larvae feed on this plant in August and September, and then pupate within flimsy silk cocoons, where they remain until the next summer.

2 The Spinach (*Lygris mellinata*). Wherever currant bushes grow in England, especially in gardens, this moth is common; but it is very local in southern Scotland and Wales, and in Ireland is known only round Dublin. The brown-chequered fringes to its wings distinguish it from the Northern Spinach. The moths fly in July and August, at dusk and after dark, and are readily attracted to light. They overwinter as eggs which hatch the following April; and the larvae feed at night on the foliage of red and black currant. They pupate in June, the light-yellow pupa, contained in a flimsy silken cocoon, lying on the ground beneath the foodplant.

3 Northern Spinach (*Lygris populata*). Unlike the Spinach, this species has chequered fringes only on the forewings. Those further north in Scotland are darker, and in the northern mountains are almost black. The females are generally smaller and paler than the males. They are very common on heaths and moors and in woods all over the British Isles, except for the south of England, but are locally plentiful in parts of the south-west, such as Exmoor. They fly at night in July and August and rest by day amongst the foodplants, whortleberry, crowberry, or sallow, on which they lay their eggs. The larvae hatch the following spring and pupate in June and July inside cocoons formed by drawing together the leaves of the foodplant with silk.

4 Barred Yellow (*Cidaria fulvata*). This moth frequents hedgerows throughout the British Isles, apart from the Shetlands; it is common in most parts of England and Wales, but local elsewhere. It flies in the evening and after dark in June and July. The dark-yellow eggs overwinter, and the following spring the larvae feed on wild and garden roses and pupate in June within delicate silk cocoons among the leaves. The pupae are a dull, light-brown in colour.

5 Blue-bordered Carpet (*Plemyria rubiginata*). A locally common moth to be seen in the evenings and after dark in July and August throughout the British Isles, especially where alders grow. The northern sub-species, *plumbata*, has an uninterrupted central band on each forewing, while another northern form has the forewings suffused with smoky brown. The eggs overwinter and hatch in the spring; and the larvae, which feed on alder, birch, blackthorn, and wild and culti-vated apple, are ready to pupate in June. The shiny brown pupa is hidden within a silken web spun up on the twigs of the foodplant.

6 Red-green Carpet (*Chloroclysta siterata*). This is a woodland insect, found throughout the British Isles, except in the extreme north, but far more plentiful in some districts than others. It differs from the Autumn Green in having darker green forewings, which have a rosy tint, and darker, grey-brown hindwings. The female is even darker and has broader wings. They are on the wing at night in the autumn, when they visit ivy blossom, hibernate as adult moths, and in the spring visit sallow catkins. The larvae feed from June to August on the foliage of trees, including oak, ash, birch, blackthorn, and are said to pupate within flimsy silk cocoons in bark crevices.

7 Autumn Green Carpet (*Chloroclysta miata*). A paler moth than the Red-green and without the rosy tint on the forewings. It flies at night in the autumn, when it comes to ivy blossom, and again in the spring, after hibernation, when it visits sallow catkins. It is generally common in wooded districts throughout the British Isles. Its larvae feed from June to August on various trees, including alder, birch, oak, and sallow. The pupae lie on the ground below the trees inside narrow, silken cocoons.

8 Common Marbled Carpet (*Dysstroma truncata*). The ground colour of the forewings of this extremely variable, widespread, and common moth range from whitish or yellowish-brown to almost black, and are often marbled with attractive markings. It closely resembles another common species, the Dark Marbled Carpet (*D. citrata*), but the Dark Marbled has an angled line on the underside of the hindwing which the Common Marbled has not. The Common Marbled Carpet flies at night in woodlands and around hedge-rows in May and June, and again, as a second gene-ration, in August and September. In some years there is a third generation in the south in December. The larvae, which hatch from shiny yellow eggs, feed on various plants, including hawthorn, sallow, birch, whortleberry, and wild strawberry. The first brood feed from September until they hibernate, and then again until they pupate in April; the second brood in June and July. The light-green pupae lie within shelters formed by drawing a leaf together with strands of silk.

LIFE SIZE

1 PRETTY CHALK CARPET   2 THE SPINACH
3 NORTHERN SPINACH   4 BARRED YELLOW
5 BLUE-BORDERED CARPET   6 RED-GREEN CARPET
7 AUTUMN GREEN CARPET   8 COMMON MARBLED CARPET

There is a note on Geometers on p. 96, and on the Carpet sub-family on p. 98.

1 **Grey Pine Carpet** (*Thera obeliscata*). The forewings of this variable moth may be almost black, and in the Scottish Highlands the cross-bands are red. It is common after dark in May and June in most parts of the British Isles where there are Scots pines, on which its larvae feed. The larvae hibernate and feed again in the spring. The pupae are contained in flimsy silken cocoons.

2 **July Highflyer** (*Hydriomena furcata*). A very common and also very variable Geometer, to be found almost everywhere in the British Isles in hedgerows and wood borders. The moths fly from late June to August, and the greenish-white eggs overwinter and hatch the following May. The larvae normally feed at night on sallow, willow, poplar, and hazel; but there is a smaller moorland race, commonest in the north, which feeds on whortleberry. The smooth, bright-brown pupae lie within cocoons spun up on the foodplant.

3 **Argent and Sable** (*Rheumaptera hastata*). There are two sub-species of this moth: a southern one, which is the typical form and illustrated here, and a northern one, *nigrescens*, locally common throughout Scotland and the surrounding islands, which is smaller and more heavily spotted on the white regions of the wings. It flies in June and July, and the larvae feed chiefly on bog myrtle and whortleberry, from July to September. The southern sub-species inhabits the rest of the British Isles, including south and west Ireland, and although plentiful where found, tends to be very local. The moths fly actively in the sunshine, usually around birch trees, in May and June, and the larvae feed on birch and sallow from June to August. Both in the north and south, the larvae form shelters by spinning together the leaves at the tips of the foodplant twigs with silk, inside which they later pupate. The moths emerge the following year.

4 **The Streak** (*Chesias legatella*). This moth has a conspicuous streak extending to the tips of its forewings; it also has rather oval-shaped wings. It flies for a short time after dark in the autumn around clumps of broom, and then settles on them for the rest of the night and for the day. It is common wherever broom grows throughout Britain; in Ireland it is confined to the north where it is local. Apart from the fact that the larvae feed on broom during the spring, little is known about its life-history.

5 **The Treble-bar** (*Anaitis plagiata*). A larger and darker moth than its very similar close relative, the Lesser Treble-bar (*A. efformata*), which is less widespread; the male also has shorter abdominal claspers. There is

a Scottish race, *scotica*, with more blue-grey forewings. The Treble-bar is generally common in open country, especially on chalk and limestone hills, almost everywhere in the British Isles, and flies in May and June, and again, as a second generation, in August and September. It often flies by day amongst St. John's Wort, on which its larvae feed. The second brood of larvae overwinter, and pupate the following April. The pupae are reddish-yellow, speckled with white.

6 **Common Carpet** (*Epirrhoë alternata*). A common and variable moth easily confused with relatives such as the Wood Carpet (*Epirrhoë rivata*). A brown, smaller race is found in the Hebrides. The Common Carpet flies in May and June and again, as a second generation, in August and September, throughout the British Isles, except for the Shetlands. The larvae feed on bedstraw, and the second brood pupate in the autumn and pass the winter as reddish-brown pupae lying within flimsy silken cocoons amongst the ground litter.

7 **November Moth** (*Oporinia dilutata*). A very familiar autumn moth but not easy to distinguish from the Pale November Moth (*Oporinia christyi*), except under a microscope. It is, however, generally darker in colour. Exceptionally dark forms are frequent in some districts. The moths fly at night in October and November, but may often be seen flitting through their woodland haunts during the day or resting on fences and tree-trunks. They are common almost everywhere in the British Isles, except the extreme north of Scotland. The larvae, which hatch from crimson eggs the following spring, feed from April to June on the foliage of many trees, including oak, birch, elm, and sallow, which they and the Winter Moth larvae often strip of leaves in late spring. The brown pupae lie in underground cells.

8 **Winter Moth** (*Operophtera brumata*). A well-known pest of fruit and other trees in the spring. The female has only rudimentary wings and cannot fly. After emerging from her pupa at the base of the tree, she makes her way up the tree-trunk and deposits her clusters of greenish-white eggs, which turn orange, in bark crevices and on unopened buds. The winged male is smaller and usually darker than the very similar Northern Winter Moth (*O. fagata*). Winter Moths are out during the winter, sometimes as early as October, and are abundant in wooded districts or orchards all over the British Isles. The larvae hatch in April and feed on the foliage of many kinds of trees and shrubs, until they pupate in May or June in the soil. Their numbers vary greatly from year to year. Recent studies of these fluctuations made near Oxford have shown, among other things, that the appearance of the larvae tends to synchronize in time and place with the breeding season of some small birds, such as tits, which feed on them.

LIFE SIZE

1 GREY PINE CARPET   2 JULY HIGHFLYER
3 ARGENT AND SABLE   4 THE STREAK
5 THE TREBLE-BAR   6 COMMON CARPET
7 NOVEMBER MOTH   8 WINTER MOTH

There is a general note on the Geometers on p. 96, and on the sub-family Larentiinae on p. 98. The Pugs are notable for their small size and habit of resting on a surface with their wings held away from the body and spread flat out. The larvae mostly feed inside flowers or seed-pods, and most species pupate in the ground and overwinter in cocoons constructed from silk and soil particles and closely resembling pieces of earth. The pupae are often brilliantly coloured.

1 **Foxglove Pug** (*Eupithecia pulchellata*). This prettily-marked little moth resembles the less widespread Toadflax Pug (*E. linariata*), but is larger and duller. It flies in May and June wherever foxgloves grow in profusion in the British Isles. Its larvae feed from June to August on the stamens and unripe seeds inside the foxglove flowers, and then close the mouths of the flowers with silk; the faded flowers remain on the plants after the others have dropped. The pupae are reddish-yellow with pale green wing-cases.

2 **Netted Pug** (*Eupithecia venosata*). A common moth which flies in May and June all over the British Isles wherever bladder, sea, and other campions grow. Its larvae feed in the seed-pods from late June until early August, when they change into bright-red pupae which overwinter.

3 **Lime-speck Pug** (*Eupithecia centaureata*). A common species which flies almost everywhere in the British Isles, except the north of Scotland, from May to August and again, as a second generation, in the autumn. The eggs, at first whitish but later light orange, are laid on knapweeds, ragwort, scabious, yarrow, and other plants, on the flowers of which the larvae feed throughout the summer. The pupae are a light red colour.

4 **Currant Pug** (*Eupithecia assimilata*). This moth flies at dusk, the first generation in May and June, and the second in August. It is locally common wherever currants and hops are to be found, on which the larvae feed in June and July, and again (second brood) in the autumn. The pupae are olive-brown.

5 **Common Pug** (*Eupithecia vulgata*). A very common and also variable Pug, to be seen almost throughout the British Isles flying in May and June, and as a second brood in August. The larvae feed on the foliage of various plants, including sallow, willow, hawthorn, and bramble in June and July, and as a second brood in the autumn. The pupae are reddish-brown and olive.

6 **Grey Pug** (*Eupithecia castigata*). A common moth which flies in May and June, and sometimes in September as well, everywhere except for the Orkneys and Shetlands. Its larvae feed from August to October upon the foliage and seeds of a great variety of plants. The pupae are chiefly greenish-yellow.

7 **Bordered Pug** (*Eupithecia succenturiata*). So named because of the clouded borders to the wings, this moth flies in July in many parts of England and Wales, being commonest on the coast; but it is scarce in Ireland, and there is only one Scottish record. The larvae feed in the autumn on mugwort, wormwood, tansy, and yarrow, and pass the winter as yellow-brown pupae, with olive-green wing-cases.

8 **Narrow-winged Pug** (*Eupithecia nanata*). A moth distinguished by the cross-lines on its pointed forewings. It is to be seen on heather moors and heaths all over the British Isles, flying in May and sometimes again in late summer. Its larvae feed from June to September on heather flowers. The pupae are reddish with yellow wing-cases.

9 **Green Pug** (*Chloroclystis rectangulata*). An extremely variable moth which may be predominantly green, brown, grey, or black. The black form is becoming increasingly common in industrial areas and widespread. It flies in June and July over most of the British Isles. The females lay their eggs on wild and cultivated apples and pears and on hawthorn and blackthorn. The larvae hatch the following spring and then feed on the flowers. The moths emerge from the red and yellow pupae in about two or three weeks.

10 **V-Pug** (*Chloroclystis coronata*). This moth can be distinguished from green forms of the Green Pug by the black V-mark on the forewings. It flies in May and, as a second generation, in August, and is common in most parts of southern England, south Wales, and Ireland; but it is scarce further north and absent north of the Scottish Lowlands. The larvae feed in June and July, and as a second brood in the autumn, on the flowers of plants such as hemp agrimony, traveller's joy, purple loosestrife, and golden-rod. The pupae are light reddish-yellow, spotted with black.

LIFE SIZE

1 FOXGLOVE PUG   2 NETTED PUG
3 LIME-SPECK PUG   4 CURRANT PUG
5 COMMON PUG   6 GREY PUG
7 BORDERED PUG   8 NARROW-WINGED PUG
9 GREEN PUG   10 V-PUG

There is a general note on Geometers on p. 96. This sub-family contains mainly medium-sized or even comparatively large moths, although some of the species are small. In some species (*see* p. 108) the females are almost or completely wingless. The larvae are generally stick- or twig-like, in some cases, remarkably so.

1 **Magpie Moth** (*Abraxas grossulariata*). The warning colours of the moths, larvae, and pupae of this species remind predators, such as birds, that they are distasteful. The slow-flying moths, which vary a good deal, are on the wing in July and August and are very common in gardens, hedgerows, woods, and heaths, all over the British Isles. The larvae hatch in late summer from batches of yellowish eggs laid beneath the leaves of the foodplant — blackthorn and hawthorn, or trees such as apple, plum, or elm, or, in the north, heaths and heather. They are also a serious pest of cultivated currants and gooseberries. The following June they change into black pupae, ringed with yellow, and contained within flimsy, transparent, silk cocoons spun under the leaves of the foodplant or in some such place.

2 **Clouded Border** (*Lomaspilis marginata*). A generally common moth in wooded districts in most parts of the British Isles, except in the extreme north, and varying a good deal in the extent of the blackish-brown markings. It flies at dusk or night in May and June, and again, as a second brood, in August, and is often to be seen during the day. The larvae feed on the foliage of sallows, willows, or aspen in June and July, and as a second brood in the early autumn. The pupae overwinter, contained in fragile silk cocoons spun-up amongst the foodplant.

3 **Common White Wave** (*Deilinia pusaria*). This species is very like the Common Wave (*Deilinia exanthemata*), except that the cross lines on the forewings are greyer and straighter. There are two generations each year in wooded districts throughout the British Isles: the first in May and June, and the second in August. The moths fly at dusk, but may often be disturbed from the vegetation by day. The larvae feed upon alder, birch, and sallow in June and July and again in August and September. They overwinter as pupae within flimsy silken cocoons amongst the ground litter.

4 **Light Emerald** (*Campaea margaritata*). The attractive pale green of the freshly emerged moth soon fades and becomes yellowish-white. It flies at night in June and July in wooded districts throughout the British Isles, except the extreme north, and is generally plentiful. The larvae feed from August until the following May on the foliage of various trees, especially oak, birch, beech, elm, and hawthorn; they generally hibernate through the winter, except during mild spells.

5 **Canary-shouldered Thorn** (*Deuteronomos alniaria*). A fairly common moth which flies at night from late August until October in wooded districts over most parts of the British Isles, except the far north. It rests on tree-trunks or fences during the day, closely resembling a shrivelled leaf. The rows of green eggs, which later become blackish, are laid on the twigs of alder, birch, or oak, and do not hatch until the following May. The larvae pupate in late July within silk cocoons on the ground amongst the undergrowth.

6 **Early Thorn** (*Selenia bilunaria*). There are two generations of this moth each year; the first appearing in April and May, and the second smaller and paler brood (6a) in July and August. There are also occasional darker (melanic) moths and larvae in some industrial districts. They are common in wooded districts throughout the British Isles, apart from the extreme north. The very twig-like larvae can be seen in May and June and again in August and September feeding on the foliage of trees and shrubs such as alder, birch, blackthorn, elm, hawthorn, lime, and sallow. They pass the winter as pupae contained within silken cocoons and lying amongst the leaf litter.

7 **Scalloped Hazel** (*Gonodontis bidentata*). This species varies considerably, particularly in the shade of the brown ground colour, which may range from pale to dark, and in industrial regions of northern England is often very dark (melanic). The moths fly at night from April to June and are abundant throughout the British Isles, except for the Orkneys and Shetlands. The eggs, which are laid in batches around the twigs of the foodplant, are bright blue at first but later change to chestnut brown. The larvae feed from late June to October on the foliage of trees such as oak, birch, sallow, larch, blackthorn, and hawthorn, and the chestnut-brown pupae overwinter inside silken cocoons on the ground beneath dead leaves and moss.

8 **Scalloped Oak** (*Crocallis elinguaria*). The shade of the ground colour or the cross-band on the forewings varies a good deal with this species, which is common almost everywhere in the British Isles, except for the extreme northern isles. It flies at night in July and August, frequently coming to light. The light-grey eggs, clouded with darker markings, are laid in rows upon many kinds of trees and shrubs. The larvae feed on the leaves from September until June, normally hibernating in the winter. The pupae are contained in silken cocoons spun-up among fallen leaves or under moss growing on the tree-trunks.

LIFE SIZE

1 MAGPIE MOTH   2 CLOUDED BORDER
3 COMMON WHITE WAVE   4 LIGHT EMERALD
5 CANARY-SHOULDERED THORN   6 EARLY THORN
7 SCALLOPED HAZEL   8 SCALLOPED OAK

There is a general note on Geometers on p. 96, and on the sub-family Ennominae on p. 106.

1 Swallow-tailed Moth (*Ourapteryx sambucaria*). A common moth in most parts of England, Wales, and Ireland, and much less frequently in southern Scotland. It flies at night in July along wood borders, hedgerows, and in large gardens. The females deposit their batches of scarlet-red eggs under the leaves of ivy, privet, oak, hawthorn, and other trees and plants, on which the larvae feed from August until the following June, hibernating in the winter in bark crevices. In June the black-spotted, pale-brown pupae are suspended hammock-fashion in flimsy cocoons beneath a leaf or twig of the foodplant.

2 Brimstone Moth (*Opisthograptis luteolata*). This moth, which is common all over the British Isles, frequents wooded districts, hedgerows, gardens, and many other habitats. It flies at night between April and September, and sometimes later. The larvae feed principally on hawthorn and blackthorn, between June and October. Many pupate before the winter and give rise to the spring moths; others hibernate and pupate in the spring. The pupae lie within thick, silken cocoons either amongst the ground litter or attached to a leaf.

3 Speckled Yellow (*Pseudopanthera macularia*). This day-flying moth may be seen fluttering through woodlands in May and June in most districts of the British Isles. It is widespread and common in the southern half of England and Wales, but only locally common in northern Wales and southern Ireland, and scarce elsewhere. The larvae feed upon wood sage, woundworts, and dead-nettles from late June until August; then they pupate within weak, silken cocoons which lie amongst the ground litter until the moths emerge the following spring. The pupae are chestnut-brown.

4 Early Moth (*Theria rupicapraria*). The male appears on the wing in the winter. Both the males and the almost wingless females may be found on mild nights in January and February sitting on the twigs of hawthorn and other hedgerow shrubs. They are common in most parts of the British Isles, scarce in southern Ireland, and absent from northern Scotland. The shiny eggs are laid in large batches on the twigs of blackthorn, hawthorn, oak, wild plum, or bilberry, upon the foliage of which the larvae feed in April and May. They pupate in late spring within slight silken cocoons on the ground beneath the foodplants.

5 Spring Usher (*Erannis leucophaearia*). This moth, as its name implies, is one of the first moths to appear in spring — in February or March. The males, which fly at night, and the almost wingless females rest during the day on clusters of dead leaves in the trees or on the tree-trunks, as well as on fences and walls. The wing-markings of the males vary considerably, and a dark (melanic) form (5a) is becoming increasingly common in the south. The larvae hatch in April from eggs laid in the bark crevices of oak trunks, and feed upon the oak foliage, joining the young leaves together with silk to make shelters. They pupate in June on the ground amongst the leaf litter, and overwinter. The Spring Usher is found in oak woods throughout Britain, apart from northern Scotland, but is common only in the south. It has only once been recorded from Ireland.

6 Dotted Border (*Erannis marginaria*). The dotted margins to all four wings of the male are not easily seen in the darker (melanic) specimens. The female has much abbreviated, functionless wings. They appear in March and April, and may be seen resting, both at night and by day, on twigs in hedges and trees, and also on tree-trunks and fences. The males fly for a brief spell at dusk. The larvae feed from April to June on the foliage of various trees, including oak, birch, sallow, and hawthorn. They pupate in the soil and emerge the following spring; the pupae are shiny brown in colour. They are common throughout the British Isles, except in the extreme north.

7 Mottled Umber (*Erannis defoliaria*). The forewings of the male vary enormously, two varieties being shown here; a dark form also is becoming increasingly frequent in some districts. The spider-like females are wingless. They deposit their yellow eggs in bark crevices, and these hatch in April or May. The larvae feed on the foliage of various trees and shrubs, such as oak, elm, birch, and cultivated fruit trees. The larvae, like some other geometer larvae, have a habit of dropping from the foliage on silk threads when alarmed, and pulling themselves up again when the danger is past. They often occur in sufficient numbers to defoliate the trees in late spring. They pupate in June amongst the ground litter, and the moths emerge from October to January, or later. They sit by day on tree-trunks and are generally common in wooded districts all over the British Isles, except for the far north.

8 Pale Brindled Beauty (*Phigalia pilosaria*). The males vary a good deal in the markings and ground colour of the forewings, and in industrial areas a dark (melanic) form is increasing (*see* p. 199). The moths appear from January to March and are generally common in most parts of Britain, except for the extreme north. They occur in Ireland but are scarce. The males fly at night, and rest by day with the wingless females on tree-trunks in wooded districts. The larvae feed from April to June on the leaves of various trees and shrubs, including oak, elm, hawthorn, and sallow. They pupate in underground cells.

LIFE SIZE

1 Swallow-tailed Moth	2 Brimstone Moth
3 Speckled Yellow	4 Early Moth
5 Spring Usher	6 Dotted Border
7 Mottled Umber	8 Pale Brindled Beauty

There is a general note on Geometers on p. 96, and on the sub-family Ennominae on p. 106.

1 Brindled Beauty (*Lycia hirtaria*). This moth is very often seen in London, especially in the Royal Parks and squares; it is fairly common in southern England, and is found locally in other parts of Britain, except the extreme north. The moths fly in March and April and rest by day on tree-trunks and fences. The wings of the female look weaker and have fewer scales than those of the male. The larvae feed from May to July upon elm, lime, sallow, willow, apple, pear, and plum, and the dark chestnut-brown pupae overwinter in underground cells.

2 Oak Beauty (*Biston strataria*). A plentiful moth in wooded districts in many parts of England and Wales; but it is scarce and local in Ireland. The moths appear between late February and early May, and rest by day low down on tree-trunks. The male, which comes freely to light, is smaller than the female and has feather-like antennae. The larvae feed from May to July chiefly on the foliage of birch, oak, and elm. The dark-brown pupae overwinter below the ground.

3 Peppered Moth (*Biston betularia*). The lighter coloured moth is the typical form, but a completely black (melanic) form, *carbonaria* (3a), is becoming more common and widespread, especially in industrial regions. An intermediate form, *insularia*, which is black, speckled with white, is also found in many districts. The females are larger than the males and have much thinner antennae. The moths fly by night in May and June, sometimes later, and are generally common in woodlands, parks, and gardens throughout the British Isles, except for the northern isles. The larvae feed from July to September on various trees and bushes, including oak, birch, elm, sallow, plum, and bramble; they pupate in the soil, where the chestnut-brown pupae pass the winter.

4 Waved Umber (*Menophra abruptaria*). The wing markings of this species are adapted to give it good camouflage when resting by day on splintered wood, such as broken tree-trunks, or on fences, or walls. A dark (melanic) form is becoming common round London. The moths fly at night in April and May and are fairly common in England and Wales, especially round London, but almost unknown in Scotland and Ireland. The females, which are generally larger and paler than the males, deposit their batches of greyish-green eggs upon the foliage of lilac and privet, on which the larvae feed from May to August. The well-camouflaged and tough cocoons, constructed of silk mixed with chewed pieces of wood, are attached to the twigs or branches of the foodplant, where they overwinter.

5 Willow Beauty (*Cleora rhomboidaria*). The ground colour of this moth varies from light brown, through the more usual greyish-brown, to dark grey. The moths are common almost everywhere in the British Isles, and fly at night in July and August, and sometimes again in September. The larvae hatch from pink eggs mottled with green, and feed on various plants, including birch, hawthorn, ivy, lilac, and privet, from August until May, and then hibernate. Some larvae feed up quickly and produce second-brood moths. The shiny-brown pupae in tough, silken cocoons are attached to twigs of the foodplant.

6 The Engrailed (*Ectropis biundulata*). This variable moth is easily confused with the Small Engrailed (*E. crepuscularia*), but it is generally darker and has two broods to the other's one. The first generation flies in March and April, and the second in July, whereas the Small Engrailed is on the wing in May and June. Both species rest by day on tree-trunks. The Engrailed is common over most of Britain and also occurs in south-west Ireland. Its larvae, which hatch from green eggs, feed on various trees, such as oak, in May and June, and again in July and August. The chestnut-brown pupae overwinter underground.

7 Common Heath (*Ematurga atomaria*). The males of this very common little moth are slightly larger and darker than the females, and have feather-like antennae; both sexes vary a great deal. They fly actively in the sunshine in May and June, and sometimes in August, on heaths and moors almost everywhere in the British Isles. The larvae feed chiefly on heaths and heather and on leguminous plants in July and August, sometimes later. They overwinter as pupae.

8 Bordered White (*Bupalus piniaria*). A common moth in many pinewoods and conifer plantations throughout the British Isles, but local in Ireland. There are two forms: in the northern race the ground colour of the male is white (8a); that of the female yellowish brown; in the larger, southern English race the ground colour is yellow in the males and orange in the females. Hybrids occur in the north Midlands, and an intermediate form is found in Ireland. The moths fly by day from May to July; their larvae feed on the foliage of Scot's pine and other conifers from July until October and are sometimes a pest. The dark-brown and green pupae overwinter in cells in the soil.

9 V-moth (*Itame wauaria*). The V-mark on each forewing gives this moth its name. It is generally common over most of the British Isles, except in Ireland where, though widespread, it is scarce. V-moths fly in July and August in places where currants and gooseberries are grown, on which the larvae are pests. They attack the leaves and young shoots from April until June, and then pupate in flimsy, silken cocoons among the leaves.

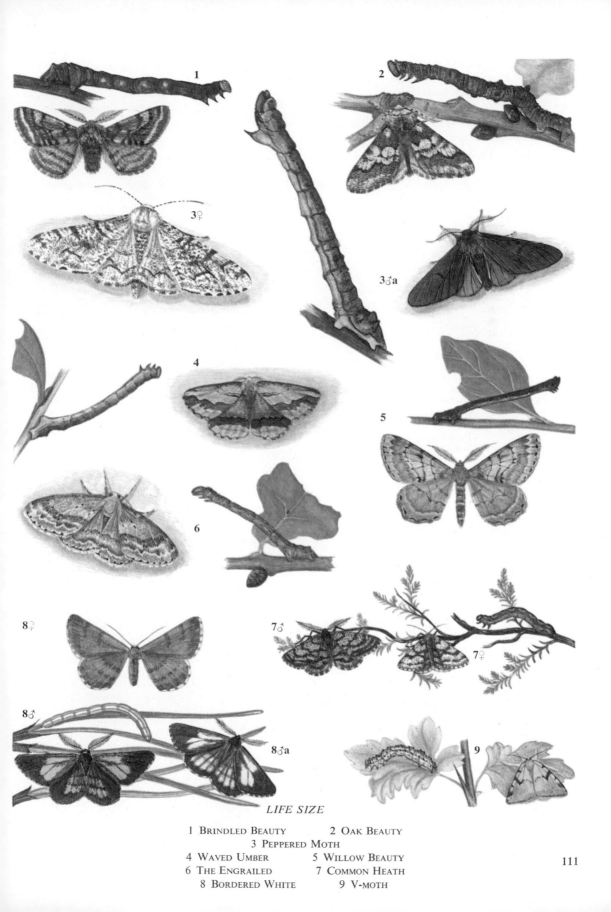

LIFE SIZE

1 Brindled Beauty   2 Oak Beauty
3 Peppered Moth
4 Waved Umber   5 Willow Beauty
6 The Engrailed   7 Common Heath
8 Bordered White   9 V-moth

The first two species on this page belong to the Geometrid sub-family Ennominae (*see* p. 106). The Goat and Leopard Moths are Cossidae — night-flying, medium to large moths, without functional tongues, whose larvae feed inside the stems of trees and other plants. The Burnets and Foresters are Zygaenidae — day-flying, usually brightly-coloured moths, which live in colonies. The bright colours of the Burnets warn predators, such as birds, that they are distasteful (*see* p. 200). The short, stout larvae construct white or yellowish, paper-like cocoons on plant stems. When the moths emerge, the empty black pupa-cases are left sticking out of the cocoons. The female Burnets have stouter bodies than the males.

1 Brown Silver-line (*Lithina chlorosata*). A generally common moth wherever bracken grows throughout the British Isles; in Scotland it is plentiful only in the south. The moths fly at dusk in May and June, and by day often sit on the bracken, on which the larvae feed in June and July. The eggs, yellowish at first, soon become red.

2 Latticed Heath (*Chiasmia clathrata*). The females of this day-flying moth are smaller than the males. There are two generations, except in Ireland — the first in April and May, the second in July and August. They are generally plentiful everywhere, except in northern Scotland. They frequent grassy places where clovers, trefoils, and other leguminous plants grow, on which the larvae feed in June and July and again in September. The chestnut-brown pupae pass the winter in the soil.

3 Goat Moth (*Cossus cossus*). A large moth which flies in June and July and is widespread, though only really common in southern England. The females are much larger and bulkier than the males. In the daytime they rest on tree-trunks or fences. The brown eggs are laid on the bark of trees such as ash, elm, poplar, and willow, and the larvae bore large tunnels into the solid wood, often causing severe damage, and making the trees exude sap, which is attractive to many insects. They are not fully grown for three or four years, and then they pupate either near the entrance to a tunnel or underground at some distance from the food-tree, Fully-fed larvae, normally to be seen in the autumn, have a repulsive smell suggestive of a goat. The large, glossy, chestnut-brown pupae are contained in tough, silk-lined cocoons.

4 Leopard Moth (*Zeuzera pyrina*). This moth, of which the female is considerably larger than the male, flies at night from June to August, and rests by day on tree-trunks. Leopard Moths are fairly plentiful in the south and east, and can be found in most parts of England, except for the northern counties. The larvae feed in the wood of trees and shrubs such as oak, ash, elm, chestnut, hawthorn, and apple. They take two or three years to mature, and then, in late spring, they build strong cocoons of silk and pieces of wood near the entrance of their burrows. The pupae are chestnut brown.

5 Five-spot Burnet (*Zygaena trifolii*). This moth has shorter forewings than its close relative the Narrow-bordered Five-spot Burnet (*Z. lonicerae*), less sharply-pointed hindwings, and the two central spots on each forewing united, giving an appearance of four spots. There are two sub-species: *palustrella* flies in May and June on downland, and the larvae feed upon bird's-foot trefoil and pupate low down in the vegetation; and the larger *decreta* flies in late July and early August over marshland, where the larvae feed upon greater bird's-foot trefoil and construct cocoons high up on plant stems. Both sub-species appear to be restricted to England, Wales, and the Isle of Man, where they are locally common. The larvae feed from late summer until May, hibernating while still small.

6 Six-spot Burnet (*Zygaena filipendulae*). The British sub-species *anglicola* is very common in many parts of England, particularly in the south, and throughout Ireland. The moths fly in sunshine from June to August in open, grassy places with plenty of flowers. The larvae, which hatch from yellow eggs laid in batches, feed on various vetches and clovers, especially bird's-foot trefoil, from August until May, hibernating during the winter. They construct their cocoons high up on slender, swaying grasses, which are difficult for birds to attack.

7 The Forester (*Procris statices*). This moth may be distinguished from the Cistus Forester (*P. geryon*) and the females of the much rarer Scarce (*P. globulariae*) by its much larger size; the males of the latter have pointed instead of thickened tips to the antennae. The female is slightly smaller than the male and has simpler antennae. The moths fly in the sunshine in June and July in meadows and on heaths where ragged robin grows, whose flowers they visit; they are locally plentiful, especially in the south, but not in the extreme north. The pale yellow eggs are laid on sorrels, on which the larvae feed from August until May, hibernating in the winter. Their strong, oval-shaped, silken, white cocoons are attached to the stems or undersides of the leaves.

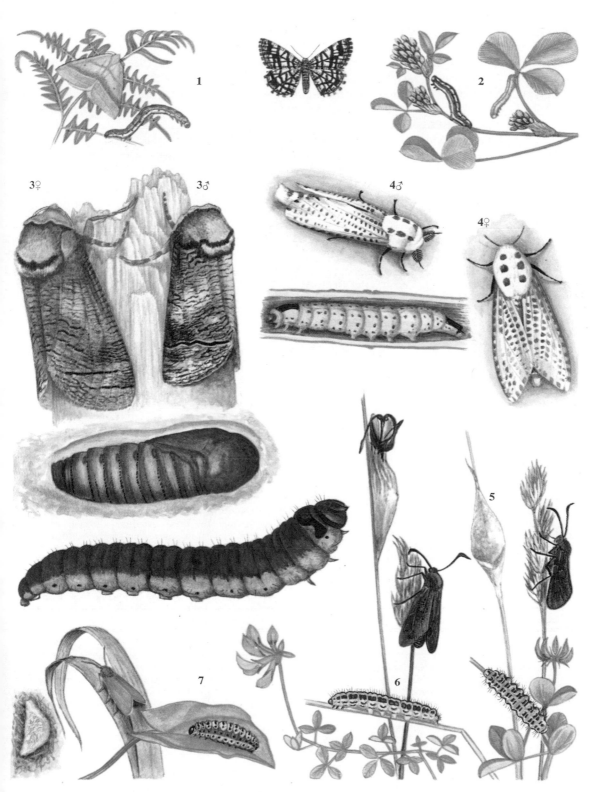

LIFE SIZE

1 Brown Silver-line   2 Latticed Heath
3 Goat Moth   4 Leopard Moth
7 The Forester   6 Six-spot Burnet   5 Five-spot Burnet

The moths on this page belong to primitive families, the first four are Clearwings, and the last three are Swifts. The Clearwings have mainly transparent wings, the forewings being long and narrow. They achieve protection from enemies by mimicking insects such as wasps and ichneumon flies. They normally fly in bright sunshine. The plant-boring, rather maggot-like larvae form pupae in cocoons near the exits from their holes.

The Swifts have no proboscis, very short antennae, and a different way of interlocking the fore and hindwings from most moths. The females scatter their numerous eggs on the ground, and the white larvae attack the roots of plants. They pupate just below the surface of the soil.

1 **Hornet Moth** (*Sesia apiformis*). This moth mimics a Hornet (*see* p. 155) very closely. It is distinguished from the Lunar Hornet Moth by its yellow instead of black head and yellow shoulder patches. The female has a rather stouter body than the male. They are not uncommon, especially in the east, and often sit on tree-trunks from May to July. They are rare and very local in Ireland, Wales, and southern Scotland, and not known in northern Scotland. The larvae burrow into the wood of aspens and poplars just above ground level, and take about three years to become full grown. They construct cocoons of gnawed wood and silk in August and change into shiny brown pupae the following spring.

2 **Lunar Hornet Moth** (*Sphecia bembeciformis*). This Hornet-mimicking moth has a distinct yellow collar, and the female is slightly the larger. They fly in June and July and are widespread and not uncommon, except in northern Scotland. The larvae feed in the trunks of poplars, sallows, and willows, taking about two years to become fully grown. The cocoons and pupae are similar to those of the Hornet Moth, but slightly smaller.

3 **Currant Clearwing** (*Aegeria tipuliformis*). A rather ichneumon-like little moth which is a pest of currant and gooseberry bushes, in the stems and shoots of which the larvae feed from August until the following May. The moths fly in June and July. The female differs from the male in having three instead of four yellow rings on its abdomen. They are found in many parts of England, Wales, southern Scotland, and Ireland, but are common only in southern England.

4 **Red-belted Clearwing** (*Aegeria myopaeformis*). An ichneumon-like moth, much like the Large Red-belted Clearwing (*A. culiciformis*) and the Red-tipped Clearwing (*A. formicaeformis*), but differing from the former in being smaller and having no light coating of reddish scales near the base of its forewings, and from the latter by having no red tips to its forewings. The female is a little larger and stouter than the male. The moths fly in July and August in gardens and orchards in many districts of England, except the northern counties. The larvae feed on the inner bark of apple and pear trees and take almost two years to become fully grown. They construct cocoons of gnawed bark and white silk in June, changing within to pale-brown pupae.

5 **Ghost Moth** (*Hepialus humuli*). The pure white upper-wings and dark underwings of the males give a ghostly effect as they swing backwards and forwards in pendulous flight over some grassy spot at dusk in June and July. The darker and larger females are attracted to them by this peculiar behaviour. In the Shetland Isles, where the nights are too light at this time of year for white wings to produce a ghostly effect, there is a local race, *thulensis* (5a), with darker, variable forewings. The larvae feed from August until the following May on the roots of plants such as dandelion, dock, and stinging nettles. The hairy pupae are brown. Ghost Moths are abundant in grassy places throughout the British Isles.

6 **Orange Swift** (*Hepialus sylvina*). The female is much larger than the male, usually with dark instead of orange-brown forewings. They fly towards dusk in late July and August in most parts of Britain where bracken grows, except for the extreme north and Ireland, but are commonest in the south. The larvae feed upon the roots of plants such as bracken and docks, from September until the following July, when they spin cocoons from silk and bits of root in cavities eaten out of the roots of the foodplant.

7 **Common Swift** (*Hepialus lupulina*). The markings on the forewings of the males vary a good deal in distinctness; the females normally have quite plain wings. The moths fly in June at dusk and are very common in open country except in Ireland where, though widespread, they are less common. The larvae feed on the roots of various grasses and other plants from late summer until the following April, when they pupate in the soil. The pupae are light chestnut-brown.

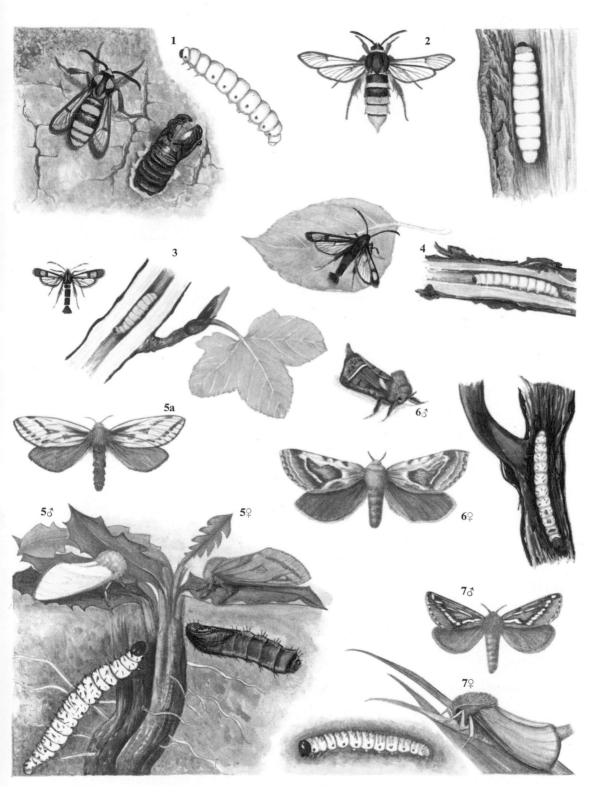

LIFE SIZE

1 Hornet Moth   2 Lunar Hornet Moth
3 Currant Clearwing   4 Red-belted Clearwing
5 Ghost Moth   6 Orange Swift
7 Common Swift

The term Microlepidoptera, or 'small moths', covers the species described on this and the next three pages. There are about 1,200 British species altogether, and only a few representative ones can be given here.

The species on this and p. 118 belong to the Pyralidae, not all of which are particularly small. They have rather narrow or angular forewings and broad, fringed hindwings, and often prominent palps (sensory mouth organs), like the Snouts (*see* p. 88). Most species fly by night, and by day if disturbed, with a weak, slow flight, and are attracted to artificial light. Few visit flowers, and some cannot feed as moths. The sexes are usually similar. The larvae mostly feed at night.

1 **Honey Moth** (*Achroia grisella*). The larvae of these moths are pests in beehives, particularly neglected ones, and may cause the bees to leave the hive. They feed on the wax, and also on dead insects and dried fruit, from August to May, and wriggle violently when disturbed. The pupae lie inside the hive encased in white, silken cocoons, and the moths, of which there are two generations, fly around the hives between June and October. The Honey Moth is common throughout England, Wales, and Ireland, and is found in some parts of Scotland.

2 **Bee Moth** (*Aphomia sociella*). A common moth almost everywhere in the British Isles; it flies from June to August, and spends the day in dense vegetation from which it is difficult to dislodge. The male differs from the female in having brown and white markings on the greenish forewings. The larvae parasitize bumble bees and wasps, infesting their nests with silken tunnels. They feed on the nest materials, disused cells, and other waste products from August to October, and finally devour the honeycomb and the bee or wasp larvae. In the autumn they spin strong, brown cocoons clustered together, usually inside the nests, and hibernate until the following spring, when they pupate.

3 **Honeycomb Moth** (*Galleria mellonella*). Another pest of beehives, less common than it used to be, but still found in many parts of England, Wales, and Ireland. The moths, which are out from June to October, behave like the smaller Honey Moth, often hiding by day under the hives. The males are much smaller and paler than the females. The larvae attack the combs of Honey Bees, feeding on the wax and spoiling the honey. In the autumn they spin tough, silk cocoons clustered together in the combs, in which they hibernate, pupating in the spring.

4 **Grass Moth** (*Crambus pascuellus*). These two Grass Moths (*see* No. 5) of the sub-family Crambinae are often called Crambids. Crambids have prominent, pointed palps, narrow forewings, and very broad hindwings. They haunt grassy places in numbers together, resting by day with their wings folded tightly around them and often head downwards, on the grass stems. At night they fly actively and are attracted to artificial lights. The females tend to be larger than the males. The larvae usually live in silken galleries low down amongst the grasses on which they feed.

This species, *Crambus pascuellus*, is common on marshy ground almost everywhere in the British Isles, flying in June and July.

5 **Grass Moth** (*Agriphila tristellus*). The common characteristics of this and other Crambids are described under *Crambus pascuellus* (No. 4). These moths are variable, some having darker forewings than others; they are found in grassy places all over the British Isles, apart from Shetland. They fly from July to September, and spend the day sitting head downwards on grass-stems. The larvae feed from September to June.

6 **Indian Meal Moth** (*Plodia interpunctella*). A pest of stored grain, seeds, dried fruit, nuts, etc., which reached Britain in the 1840s and is now numerous in warehouses in many localities. The larvae foul the stores with their droppings and the silk webs they spin as they feed. They may take up to two years to become fully grown, depending on temperature, but normally they hatch in June or July, spin flimsy cocoons in crevices in the autumn, and pupate the following spring. The moths are most frequent in June and July, but are to be seen up to October.

7 **Tobacco (Cacao) Moth** (*Ephestia elutella*). Another serious pest of stored foodstuffs in warehouses in many parts of the British Isles, and probably originally introduced from overseas. The moths are most frequent in July and August but may be seen from June to October. They avoid light and hide during the daytime. The larvae feed on tobacco, cacao, and all kinds of stored foodstuffs and vegetable materials from July to April, usually taking about three months, depending on temperature, to become fully grown. They hibernate in silk cocoons, sometimes as pupae, but they usually pupate in the spring.

8 **Mediterranean Flour Moth** (*Ephestia kühniella*). This serious pest of flour-mills, stores of grain and flour, and bakehouses throughout the country reached Britain in the late 19th century. Three generations of moths fly between April and October. The larvae live in silken tubes amongst the flour, sometimes so abundantly that they mat the flour together, clogging the mill elevators and chutes. The third brood hibernate full-grown in silk cocoons and pupate in the spring.

LIFE SIZE

1 HONEY MOTH
3 HONEYCOMB MOTH
5 GRASS MOTH (*Agriphila*)
7 TOBACCO (CACAO) MOTH

2 BEE MOTH
4 GRASS MOTH (*Crambus*)
6 INDIAN MEAL MOTH
8 MEDITERRANEAN FLOUR MOTH

There is a note on Pyralidae on p. 116.

1 **Large Tabby** (*Aglossa pinguinalis*). A relatively large and generally common Pyralid which varies considerably in size and colour pattern. It flies in June and July in barns and warehouses, where it hides in dark corners during the day, and if disturbed, instead of flying, scuttles rapidly to safety. The larvae inhabit long silken tubes amongst hay, cereal, chaff, or dry seeds, on which they feed from August until May. They pupate within tough cocoons of grey silk covered with debris, and sometimes take almost two years to mature.

2 **Gold Fringe** (*Hypsopygia costalis*). A locally plentiful moth as far north as Yorkshire and Lancashire. It flies around hedgerows, old haystacks, and thatched buildings after dark from June to August, and sometimes in October, and when disturbed in the daytime, soon takes cover again. The larvae feed from August to May on hay, straw, and other dry plant matter, and frequent haystacks. They pupate in silken cocoons.

3 **Meal Moth** (*Pyralis farinalis*). A common moth in mills and places where grain is stored, throughout the British Isles. It flies from June to August, resting by day on walls with outspread wings and raised abdomen; when prodded it runs instead of flying. Its larvae feed from the safety of long silken tubes, anchored at one end, on flour and cereal refuse. They normally feed from August until they pupate in May inside flour-covered, white, silken cocoons; they may take almost two years to mature.

4 **The Brown Grey** (*Scoparia ambigualis*). There are two races of this very common moth: *ambigualis* inhabits wooded districts in lowland regions, especially in the south, and flies in May and June, spending the daytime upon tree-trunks and fences; *atomalis*, the smaller and darker race, frequents upland moors and heaths, particularly in the north and west, and flies in July and August, resting by day on heather and rocks. All are quick to take flight when alarmed. The larvae are said to feed on mosses, but little is yet known about them.

5 **Brown China Mark** (*Nymphula nympheata*). A variable moth, a very dark form of which is found in the New Forest. The moths fly from June to August and are common almost everywhere near slow-flowing or stagnant water. They are active after dark and easily disturbed from their daytime refuges in the waterside vegetation. The eggs are laid in batches underneath the leaves of aquatic plants such as pondweed, bur-reed, and frog-bit, on which the larvae feed from July until the following June. They make floating, flat, oval shelters from leaf fragments, from which they attack the undersides of the leaves. They pupate within whitish silk cocoons attached between leaves to a plant just above the water.

6 **Garden Pebble** (*Evergestis forficalis*). A familiar insect of gardens and cultivated land, where the larva is a minor pest, all over the British Isles, except the extreme north. At least two generations each year appear in May and June and again in August and September. The eggs are laid in small batches under the leaves of crops such as cabbages and turnips, on which the larvae feed in June and July and again in the autumn, eating out the centres of cabbages. The larvae of the second brood hibernate underground in tough, silk cocoons, and pupate in the spring.

7 **Mother-of-Pearl** (*Pleuroptya ruralis*). Another common moth throughout the British Isles, which flies in July, and can be easily dislodged from amongst stinging nettles and other vegetation during the day. The larvae feed from inside a nettle leaf, which they roll up with silk, in the late summer and autumn, and again in the spring. They pupate in June inside similar silk-lined shelters.

8 **Rush Veneer** (*Nomophila noctuella*). A usually common and sometimes abundant migrant throughout the British Isles, arriving in almost any month of the year. They are usually seen from June to September, when they fly at night and are readily attracted to light. The larvae usually feed from August to October on plants such as knotgrass and clover, and they wriggle furiously when touched. They pupate within tough, white, silken cocoons concealed amongst the food-plant or on the ground; some survive the winter, producing moths the following spring.

9 **Small Magpie** (*Eurrhypara hortulata*). Though of quite a different family, this moth is rather like a small Magpie Moth (p. 107). It inhabits England, Wales, and southern Scotland, being local in Scotland but very common in the south and in Ireland. It flies at night in June and July with a rather lazy, fluttering flight, and rests by day in the undergrowth. The larvae feed in August and September on stinging nettle and plants such as mint and woundwort, rolling and spinning the leaves into shelters. They pass the winter in transparent, silken cocoons placed under bark or such places, and pupate in the spring.

10 **Common Crimson and Gold** (*Pyrausta purpuralis*). This moth, rather like the Gold Fringe (No. 2), but without the gold wing borders, flies briskly, both visiting flowers in the sunshine and flying after dark, in May and June and again in July and August. It frequents open country and wood borders up to 2,000 feet almost everywhere in the British Isles. The larvae spin two mint or wild thyme leaves together and feed inside them in June, and again from July to September.

LIFE SIZE

1 LARGE TOBBY   2 GOLD FRINGE
3 MEAL MOTH   4 THE BROWN GREY
5 BROWN CHINA MARK   6 GARDEN PEBBLE
7 MOTHER-OF-PEARL   8 RUSH VENEER
9 SMALL MAGPIE   10 COMMON CRIMSON AND GOLD

119

All the species on this page belong to the Microlepidoptera group, about which there is a note on p. 116.

1 **Triangle Plume** (*Platyptilia gonodactyla*). The 'Plume Moths' have very narrow, feather-like wings and long, spurred legs. The forewings of most species are almost split in two, and the hindwings into three, by deep clefts. Their flight is weak, and when resting, they hold the wings at right angles to the body.
The Triangle Plume has a variable dark blotch on each forewing. It is common wherever its foodplant, colts-foot, grows, except in the Shetlands, and it flies actively by night in May and June, and as a second generation in August and September. It rests by day concealed in the ground vegetation. The larvae feed from June to August beneath the coltsfoot leaves, where they pupate under a web of silk and produce a second brood. These hibernate in the stems, and in the spring feed in the flowers and pupate within the seed-heads.

2 **Large White Plume** (*Pterophorus pentadactyla*). A common Plume Moth in hedgerows and similar places throughout England, Wales, and Ireland; it flies at night in June and July and also, if disturbed, in the daytime. The hairy larvae feed on bindweeds from August until they hibernate when still young. In the spring they eat the young leaves and flowers, and pupate about May under leaves.

3 **Many-plume Moth** (*Alucita hexadactyla*). The only British representative of the family Alucitidae is named because of its multi-plumed wings. The moths fly at night from August until the autumn when they hibernate under cover, such as in ivy or in barns. In the spring they are common wherever honeysuckle grows. The larvae feed in June and July within the leaves and flowers of honeysuckle and pupate on or just below ground inside strong, silken cocoons.

4 **Green Oak-roller** (*Tortrix viridana*). This moth rather resembles the Cream-bordered Green Pea (p. 87), though it is not related. It belongs to the family Tortricidae. It is often abundant where oaks grow, especially in the south, and flies at night in June and July. It is easily dislodged by day from amongst the oak foliage by jarring the branches. The eggs are deposited in small batches on the twigs but do not hatch until the following May. The larvae feed until June inside rolled-up oak leaves or, occasionally, beech, and then pupate within a folded leaf. They are one of the main oak defoliators.

5 **Geoffroy's Tubic** (*Alabonia geoffrella*). This and No. 6 belong to the family Oecophoridae. It flies with a rising and falling motion on sunny days in May and June, and is common round hedgerows and wood borders in England and southern Ireland. The long,

thin larvae probably hatch in late summer, hibernate, and feed in rotting wood from February to April.

6 **White-shouldered House Moth** (*Endrosis sarcitrella*). A common scavenger in houses, warehouses, and farm buildings, throughout the British Isles. It flies at night throughout the summer and often rests by day on house windows. The larvae feed at any time of year on almost any kind of vegetable matter or refuse and in the nests of birds, mice, and squirrels. They live in silken galleries amongst their food. Pupae are to be found between February and September.

7 **Ypsolophus harpella.** A common moth, with hook-tipped wings, in July and August, in wooded districts everywhere except northern Scotland. The eggs overwinter, and the larvae feed in May and June within flimsy webs spun beneath honeysuckle leaves, and are lively if interfered with. They pupate within slender, white cocoons attached to the undersides of leaves or amongst ground leaf-litter.

8 **Common Clothes Moth** (*Tineola bisselliella*). This well-known pest (family Tineidae) is found indoors in most parts of the British Isles, flying from June to September. Those seen are usually males, the females being more sedentary. The larvae feed within silken galleries from October to June on woollen materials, fur, and hair, and then pupate inside silken cocoons spun-up amongst their food.

9 **Raspberry Moth** (*Lampronia rubiella*). There are twenty-five British species of the family Incurvariidae, to which this pest of raspberries belongs. It flies in June where there are raspberries, being especially active in the late afternoon. The larvae feed in June and July inside raspberry or loganberry flowers, and then drop to the ground and hibernate in white cocoons amongst the leaf litter. Next spring they climb a stem of the plant, penetrate a shoot, and eventually pupate in May within cocoons. They are common except in northern Scotland.

10 **Green Longhorn** (*Adela viridella*). A very common member of the family Incurvariidae in woods in most parts of England. The moths, especially the males, have extremely long antennae. The males dance in small swarms around trees and shrubs on sunny days from April to June, while the females at first hide in the foliage and then fly out to mate. The larvae feed from June until the following March; they mine the leaves or flowers of oak, birch, and other trees, and then drop to the ground with the leaves, make flat, portable cases of leaf fragments, and from within these feed on the fallen leaves. They pupate in the spring within the cases, the males coiling their long antennae around the tips of the pupae.

ONE-AND-A-HALF LIFE SIZE

1 TRIANGLE PLUME
2 LARGE WHITE PLUME
3 MANY-PLUME MOTH
4 GREEN OAK-ROLLER
5 GEOFFREY'S TUBIC
6 WHITE-SHOULDERED HOUSE MOTH
7 YPSOLOPHUS HARPELLA
8 COMMON CLOTHES MOTH
9 RASPBERRY MOTH
10 GREEN LONGHORN

CRANE-FLIES, GNATS,
AND MOSQUITOES

1-4 Crane-flies (family Tipulidae). There are 291 species of Crane-flies in Britain, four of which are shown here. They have thin bodies, narrow wings, and such long, fragile legs that they are often called 'Daddy-long-legs'. The females of some species have no or greatly reduced wings. The greyish-brown larvae of some of them live in the soil where they destroy the roots of grasses, cereals, and other plants. These have tough skins and are commonly called 'leatherjackets'.

Tipula paludosa (1) is a very common Crane-fly, especially in meadows and pastureland. The adults normally fly after dark in the summer and autumn, when they are often attracted indoors by artificial light. By day they are easily disturbed from the grass or similar places, when they make off with a low, dancing flight, their long legs dangling beneath them. The females deposit hundreds of small, black eggs in the soil, from which the larvae hatch some two weeks later and start attacking plant roots. They feed right through the winter and become full-grown the following spring. They pupate in the earth, and some two weeks later, the slim, horned pupae push half-way out of the soil for the adults to emerge. This life history is typical of most soil-dwelling Crane-fly larvae.

Tipula maxima (2) is the largest British Crane-fly, with the greatest wingspan of all British flies. It is plentiful from April to August in the vicinity of lakes, ponds, rivers, and streams, and its larvae live at the water's edge and feed on semi-aquatic plants. They pupate in drier soil nearby.

Nephrotoma maculata (3), one of several British black and yellow Crane-flies, has down-curved black stripes on the sides of the thorax and virtually no dark markings on its wings. It is a very common pest of gardens and farms throughout the British Isles, and the adults fly from May until September.

Tanyptera atrata (4) varies somewhat in colour, the abdomen varying from black, through yellow, to red in both sexes. The males have comb- or feather-like antennae, and both sexes, which are common from April to June in most damp woodlands, have less fragile legs than most other Crane-flies. The larvae feed and pupate in damp decayed wood of deciduous trees.

5 Winter Gnat (*Trichocera relegationis*). There are ten British Winter Gnats, which belong to the family Trichoceridae and look like small Crane-flies. They get their English name from their way of 'dancing' in large swarms on winter afternoons, even when snow has fallen. This species is on the wing in most months of the year, and is generally abundant almost everywhere. They do not bite man and are quite harmless. The larvae feed on decaying leaves, fungi, and other rotting plants on the ground, and pupate in the soil.

6-8 Mosquitoes (family Culicidae). There are fifty British species of this family, many of them called Gnats. They are thin flies with long, fragile legs and narrow wings. The Mosquitoes have well-developed mouthparts, and the adults suck the nectar of flowers with their long, needle-like probosces. Only the females, however, pierce the skins of mammals and suck blood. The males differ from the females in having long palps and feather-like antennae.

The Spotted Gnat (*Anopheles maculipennis*) (6) is an example of the type of Mosquito which rests with its abdomen raised at an angle and its head in line with the rest of its body. It has spotted wings, and one subspecies (*atroparvus*) breeds in brackish water in coastal districts, the other (*messeae*) in fresh water inland. *Atroparvus* is capable of carrying the British form of malaria called ague. Spotted Gnats mate in the autumn and soon afterwards the male dies. The female hibernates through the winter in outhouses and cellars, and in the spring she deposits her floating eggs singly in stagnant water. The active larvae, on hatching, live suspended in the water in a horizontal position just below the surface, feeding on tiny food particles. The comma-like pupae also float at the surface of the water and are just as active as the larvae. The adults emerge during the summer and are common in most parts of the British Isles, flying both by day and night.

The Common Gnat (*Culex pipiens*) (7) is a very abundant Mosquito near stagnant water and is an example of the type which rests with its abdomen parallel with the surface on which it is resting and its head lowered at an angle. Its life history is similar to that of Spotted Gnats, except that the eggs are laid in raft-like batches on the water surface, in which the larvae feed with head held downwards. The adult females, which hibernate in all kinds of buildings, never suck the blood of man, only birds.

Theobaldia annulata (8) is one of the largest and fiercest Mosquitoes. It has spotted wings and white-ringed legs and body. The females attack man and other mammals, and hibernate in buildings, sometimes waking during the winter to take a blood meal. It is common and rests in the same position and has a very similar life history to that of the Common Gnat.

TWICE LIFE SIZE

1 TIPULA PALUDOSA   2 TIPULA MAXIMA
3 NEPHROTOMA MACULATA   4 TANYPTERA ATRATA
5 TRICHOCERA RELEGATIONIS   6 ANOPHELES MACULIPENNIS
7 CULEX PIPIENS   8 THEOBALDIA ANNULATA

1 Chaoborus crystallinus. A harmless and common Gnat, belonging to the same family (Culicidae) as the Mosquitoes (*see* p. 122), but it has much shorter mouthparts, quite unsuitable for blood-sucking. It is on the wing from April to September, and the larvae overwinter. These are so transparent that their internal organs can be seen quite clearly — hence their name Phantom or Ghost Larvae. They feed on small creatures such as water fleas, for which they lie in wait, suspended horizontally in the water.

2 Non-biting Midges (family Chironomidae). There are more than 380 British species, many of which are difficult to distinguish. *Chironomus plumosus*, the one illustrated, is one of the largest, and is common from April to September. These Midges resemble Gnats, but are more fragile-looking and have a hump-backed appearance. They have poorly developed mouthparts, and the males, which have plumed antennae, often swarm over trees, bushes, buildings, or water, generally towards dusk, and dance up and down. The females, attracted by these displays, fly up, and they mate in the air. They can be seen at any time of year.

The larvae of some species are aquatic; others live in moss, dung, rotting wood, or other decomposing plant materials. The wriggling larvae of *C. plumosus* are the well-known 'bloodworms' of the muddy bottoms of stagnant ponds and ditches. They feed on decaying vegetable matter.

3 Fungus Gnat (*Dynatosoma fuscicorne*). There are over 470 species of Fungus Gnats (family Mycetophilidae), of which this is one. Like the last species, they have a delicate and hump-backed appearance and thread-like antennae. The worm-like, whitish larvae of some species feed in dung, rotting wood, and other plant matter, and others, as their name implies, on various kinds of fungi. Those of this species feed in bracket fungi on trees and pupate underground. Adults are seen quite commonly from March to August.

4 Fever Fly (*Dilophus febrilis*). This and the next species both belong to a family (Bibionidae) of day-flying, mostly black, hairy insects, of which there are eighteen British species. The antennae of the males are attached below the very large eyes, which adjoin each other and are divided into two distinct regions, like those of Whirligig Beetles (*see* p. 169). The females have smaller eyes which do not touch each other. The larvae live in the soil, feeding on decaying plants or the roots of living ones, and they pupate in underground cells. The Fever Fly does not cause fever. It is on the wing almost everywhere from early spring, when it is especially abundant, until the autumn. Like St. Mark's Fly, the males engage in aerial courtship dances above the darker-winged females, which sit in the grass. The adults help to pollinate flowers, but the larvae often attack the roots of crops.

5 St. Mark's Fly (*Bibio marci*). This close relation of the Fever Fly is much the biggest of the family and is so named because it appears on the wing on or about St. Mark's Day in the spring. It is common in late April and May in grassy meadows and fields, especially in the south. Swarms of the males fly sluggishly up and down in the sunshine with dangling legs, looking for the females which are sitting not far below in the grass. When the females take wing they are seized by the males, and mating takes place in the air.

6-7 Biting Midges (family Ceratopogonidae). There are over 130 British species, all extremely small. The females, which lack the males' plumed antennae, have biting mouthparts and suck, according to the species, the blood of mammals, birds, insects, or other animals. The worm-like larvae of some species are aquatic, living in ditches or ponds rich in algae; others live on land in decomposing plants, manure heaps, or under bark. The females of *Culicoides pulicaris* (6) attack mammals, including man, especially towards dusk. Its aquatic larvae hatch in late summer, and overwinter. The adults emerge from free-swimming pupae from April to June, and again in August and September.

The females of one common Biting Midge, *Forcipomyia bipunctata* (7), suck the blood of caterpillars. The larvae live in decomposing plants and manure heaps, and the adults fly from June to October.

8 Black Flies (family Simuliidae). The females of these small, stout flies suck the blood of birds and mammals, including man. They swarm around the heads of domestic animals, and, in some parts of the world, even cause death. The females of *Simulium equinum* attack horses and cattle, biting chiefly inside their ears; the males, like other Black Flies, confine their attentions to nectar. *S. equinum* is common near rivers in lowland regions, and is on the wing from March to November. The gregarious larvae are aquatic, feeding amongst water plants in moderately-flowing water. They pupate in the water in silken cocoons, and emerge in swarms, the males hovering above the water surface, waiting for the females.

9 Owl Midges (family Psychodidae). There are 73 species of these tiny flies, which are also called Moth Flies. Their wings and bodies are covered with long hairs. They fly by day and night throughout the spring and summer, often in great numbers, and can be seen indoors on windows or on tree-trunks, particularly near water. Their tiny, whitish larvae feed on animal excrement or rotting vegetation. Some, such as *Psychoda alternata*, frequent sewage farms.

10 Scatopse notata. A very common member of the family Scatopsidae. The larvae feed on dung and rotting plants and animals, and the adults fly from March to June and again from August to October.

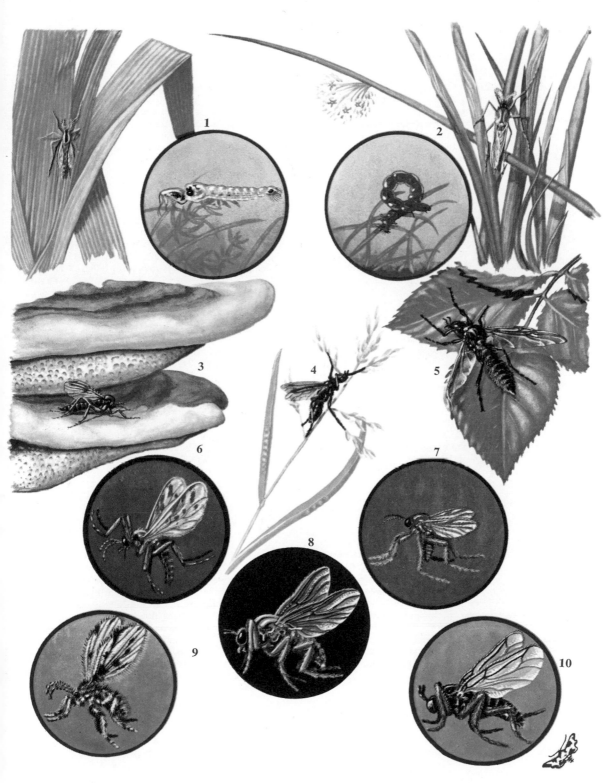

TWICE LIFE SIZE (IN CIRCLES × 6)

1 CHAOBORUS CRYSTALLINUS   2 CHIRONOMUS PLUMOSUS
3 DYNATOSOMA FUSCICORNE   4 DILOPHUS FEBRILIS   5 BIBIO MARCI
6 CULICOIDES PULICARIS   8 SIMULIUM EQUINUM   7 FORCIPOMYIA BIPUNCTATA
9 PSYCHODA ALTERNATA   10 SCATOPSE NOTATA

SOLDIER AND HORSE FLIES, ETC.

1-4 Soldier Flies (family Stratiomyidae). There are about 53 British species of these rather inactive, day-flying insects, many of which are small. They mainly inhabit damp or wet places where there are plants of the parsley family. Their name comes from their often metallic colours and their habit of resting with their wings folded over their backs so that the outer margins lie parallel to each other in stiff military fashion. Most species lay their eggs on the water surface or on water plants, and the larvae are aquatic, often floating with their tails up to the surface and feeding on other tiny water creatures. When they pupate, they retain the larval skin as a protection or a float.

Chloromyia formosa (1) is an example of a less aquatic species, the adult being commonly seen resting on garden plants from May to July; the larvae live in garden mould, rotten wood, or similar places. The female has a blue-green abdomen with a violet sheen, instead of the coppery-green sheen of the male. Members of this group do not have the twin spines on the hind part of the thorax, which are present in *Stratiomys potamida* (2), one of the largest of the Soldier Flies. This species flies with a wasp-like hum in marshes from June to August, and is locally plentiful in southern England. It differs from its near relatives in that the yellow bands on its abdomen are often unbroken, but these are less conspicuous in the female. The larvae are aquatic. *Odontomyia viridula* (3) is another insect of waterside vegetation with aquatic larvae. The abdomen of both sexes varies from white to yellow, orange, or green. These flies are very common from June to August, especially in southern England. The small wasp-like *Oxycera pulchella* (4) is also locally common in the south, and flies in July.

5 Snipe Fly (*Rhagio scolopacea*). The males of this rather large fly sit head downwards on tree-trunks and so are often called 'Down-lookers'. The name 'Snipe' refers to the sudden, darting, snipe-like flights they make from and back to their posts. There are 18 British species in the family, mostly slim, day-flying insects with long legs. *R. scolopacea*, which differs from its close relative *R. tringaria* in having spotted instead of clear wings, is common from May to July in wooded districts. How the adult feeds is still not known, but the long, whitish larvae feed on the larvae of other insects in ground litter or rotting wood, where they pupate.

6-9 Horse Flies (family Tabanidae). Most of the 28 British species are large and have several popular names, including Clegs, Stouts, and Gad Flies. They are swift, day-flying blood-suckers, but, in fact, only the females suck blood. These attack mammals, especially horses, cattle, and humans, and inflict painful bites with their strong, piercing mouthparts. The less noticeable males frequent tree-trunks and plants, sucking nectar from flowers. Many have lovely iridescent eyes. The larvae attack other larvae, small worms, etc. in moist earth and rotting wood, and suck them dry. The eggs are laid in masses on water-plants or in marshy places.

Haematopota pluvialis (6) and its three near relatives are known as Clegs and are all greyish-brown with mottled wings. The females attack silently, often not being noticed till after they have bitten. This species and *H. crassicornis*, which is commoner in the north and has darker wings and completely black antennae, are on the wing from May to September. *Chrysops relictus* (7) has iridescent green eyes, and the female has orange patches on the dark abdomen. She flies to the attack with a low hum. This fly is locally fairly common, flying in marshes and near water from June to August, and the larvae live in mud or actually in the water.

Tabanus sudeticus (8) is the bulkiest of all British flies, with a wingspan of almost 2 inches and distinctive scimitar-shaped antennae. The female of this Horse Fly attacks with a loud hum and is very persistent. It is widespread but local, particularly frequenting old forests such as the New Forest, where it flies throughout the summer. The larvae live in the soil. *Tabanus bromius* (9) is smaller and duller. It is common in southern England, flying from June to August, particularly near cattle which the females attack relentlessly. The larvae, like those of *T. sudeticus*, live and pupate in the soil.

10 Common Bee-fly (*Bombylius major*). The 12 species of British Bee-flies are all hairy, with long, hair-like legs and a very long, slender proboscis. They resemble, and some parasitize, bees. *B. major* mimics a Bumble-bee, and it hovers in the sunshine in front of primroses and other spring flowers, as Bee Hawk-moths do (p. 62), sucking nectar. It is common, flying very rapidly, in early spring in southern England, but rare elsewhere. The females scatter their eggs close to the nests of mining bees, and the larvae attack the bee larvae, sucking them dry.

ONE-AND-A-HALF LIFE SIZE

1 CHLOROMYIA FORMOSA   2 STRATIOMYS POTAMIDA
3 ODONTOMYIA VIRIDULA   4 OXYCERA PULCHELLA   5 RHAGIO SCOLOPACEA
6 HAEMATOPOTA PLUVIALIS   7 CHRYSOPS RELICTUS   8 TABANUS SUDETICUS
10 BOMBYLIUS MAJOR   9 TABANUS BROMIUS

1 Window Fly (*Scenopinus fenestralis*). A rather inactive, drab-looking fly belonging to a family of small, black flies, with only three British species. It is fairly common during the summer in southern England, usually to be seen resting on the windows of old buildings and outhouses. Its carnivorous larvae live among old carpets, rags, and other rubbish, preying upon the larvae of Clothes Moths, Fleas, and other insects.

2-5 Robber Flies (family Asilidae). All the 27 British species are day-flying, strong, hairy flies with 'bearded' faces. The mouthparts of both sexes are adapted, like those of the female Horse Flies, for piercing; but instead of sucking blood, they attack other insects, sucking them dry. The Robber Fly crouches on a leaf, twig, or on the ground until a victim passes within reach; then it makes a flying pounce, catches the prey with its bristled legs, and kills it instantly by an injection of poison from its proboscis. Some species capture their prey in flight. The eggs are laid on soil, sand, dead leaves, rotting wood, and dung, and the larvae generally feed on decomposing vegetable matter or rotting wood.

Asilus crabroniformis (2), the largest British Robber Fly, is fairly plentiful from July to September on chalk and limestone hills, heaths, pastures, and pinewood clearings in southern England. It attacks ferociously and can overcome quite large insects such as beetles, bees, wasps, and grasshoppers. The rather similar but smaller and duller *Machimus atricapillus* (3) is also fairly common in southern England and Wales and flies from June to September in wooded districts. *Laphria marginata* (4) is a striking, fairly large Robber Fly which frequents broad-leaved woods in southern England from June to August, but is not common. *Dioctria rufipes* (5), however, is very common from May to July in meadows and neighbouring hedgerows in most parts of Britain. It preys to a large extent on Ichneumon Flies and other Hymenoptera.

6-7 Empid Flies (family Empididae). There are more than 300 British species of this family, all small or medium-sized, day-flying flies with prosboces adapted in both sexes for piercing and sucking, like those of the Robber Flies. They prey on other insects, catching them in flight and sucking their juices; they also suck the nectar of flowers. The males of many species capture an insect, offer it to a female, and then pair with her while she is engaged in draining it. The males of some species of Hilara wrap up the prey in silk secreted by glands on the front legs, and one species merely offers the female an empty bundle of silk. Little is known about the larvae of Empids, except that some are predatory and live in the soil, rotting wood, or vegetable matter, while others are aquatic or semi-aquatic.

Empis tessellata (6) is common from May to August in many parts of the British Isles, being attracted to the flowers of hawthorns and umbelliferous plants. It has a leisurely flight, but is able to catch quite large insects. Males of *Hilara maura* (7), which is also common over the British Isles, have greatly dilated lower parts of the forelegs. Swarms of males perform rapid aerial evolutions throughout the day during summer over ponds and other sheets of water, or sometimes over land. These are connected with mating, which takes place in the air, the female grasping all the while the silk-wrapped insect which has been presented to her by her partner.

8- Long-headed Flies (family Dolichopodidae). The 250
10 or so British species are generally small, predacious, metallic bluish or greenish, long-legged flies, with distinctive head profiles; they also have a characteristic habit of resting with the head and thorax raised. The males of many species have large, conspicuous genitalia, and noticeable modifications to their feet. Most species are on the wing by day during the summer or early autumn, and are found near standing water or in marshy places. They prey on small insects or their larvae, crushing them in their large mouthparts. The larvae of most long-headed flies dwell in wet mud or sand, or rotting plants and wood, some even in water, and are thought to be carnivorous; but little is known of their life-histories.

Dolichopus ungulatus (8) is abundant in the British Isles near water, such as pond margins, usually in large numbers. The distinctive genitalia of the male are bent back underneath the abdomen. *Poecilobothrus nobilitatus* (9) is also abundant in many parts of the British Isles, often in swarms at the muddy edges of ditches and ponds, or on the surface of ponds overgrown with duckweed. The flies can run swiftly over the mud or weeds, and the males, which have white tips to their wings, take frequent short flights. If they land by another male, they fight briefly; if it is a female, they display with much wing-waving. The males of *Argyra diaphana* (10), another common long-headed fly, can be seen flying to and fro near the ground, their bodies alternately flashing in the sunshine and then disappearing — an effect due to a silvery sheen on their otherwise dark abdomens.

11 Megaselia rufipes. This tiny, very active fly belongs to the family Phoridae, which runs in a characteristic scuttling manner, rarely flying, and of which there are more than 250 British species, all of them very small. This species is common throughout the British Isles near decomposing animal or vegetable matter, either indoors or outdoors. It is often seen on windows, even in winter. The larvae feed on carrion or rotting vegetable matter, though not much is yet known about their life histories.

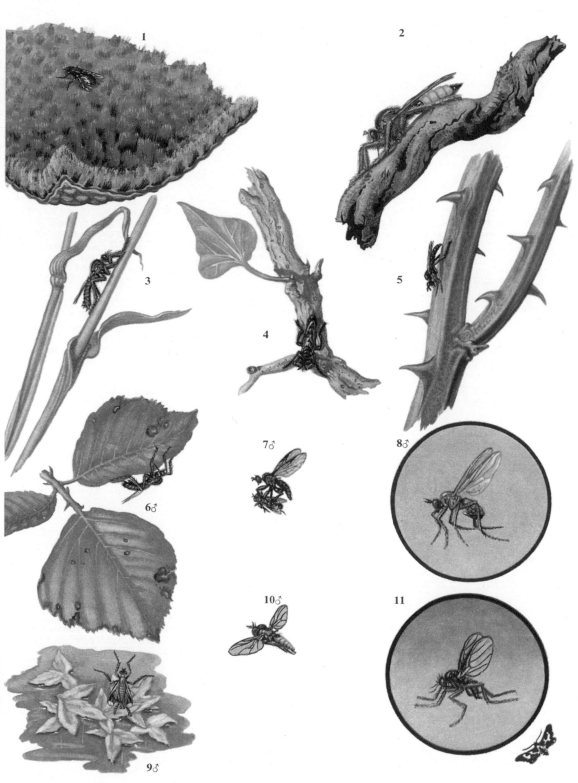

LIFE SIZE (*IN CIRCLES Fig.* 8 × 3, *Fig.* 11 × 5)

1 SCENOPINUS FENESTRALIS   2 ASILUS CRABRONIFORMIS
3 MACHIMUS ATRICAPILLUS   4 LAPHRIA MARGINATA   5 DIOCTRIA RUFIPES
6 EMPIS TESSELLATA   7 HILARA MAURA   8 DOLICHOPUS UNGULATUS
9 POECILOBOTHRUS NOBILITATUS   10 ARGYRA DIAPHANA   11 MEGASELIA RUFIPES

129

All the species on this page, which have no common names, and the first five species on page 129 belong to one family, of which there are 236 species in Britain. The males have the habit of hovering while awaiting the females, and some species make a humming or droning sound while doing so. Most adults visit flowers for nectar or pollen, and are sometimes called Flower-flies. Many lay their eggs among greenflies (aphides) on which the larvae feed; other larvae live in mud, dung, or even water, and some act as scavengers in the nests of bees, ants, and wasps.

1 **Syrphus ribesii.** A very common yellow and black Hover-fly, to be seen on the wing from April to November, feeding on nectar and pollen. Blind larvae hatch from white, sausage-shaped eggs and immediately attack aphides, a single larva eating as many as fifty a day. The larva pierces the aphis with its hooked mouth, sucks it dry, then bends backwards, leaving the empty aphis skin behind, and moves on to the next one. The larvae are so transparent that their digestive tubes can be seen. When they pupate, the brown pupae remain on the plant.

2 **Syrphus balteatus.** This Hover-fly has a similar life history to *S. ribesii*. It has been seen throughout the year, but is most abundant in the summer. The winter specimens are usually darker than the summer ones. These insects have been seen with other insects migrating through mountain passes in the Alps and Pyrenees.

3 **Scaeva pyrastri.** A common Hover-fly, widely distributed from May to November; large numbers migrate to Britain from the continent, many invading even central London. It lays its eggs among aphides, on which the larvae feed when they hatch. There is a black form (*unicolor*) without the white half-moons on the abdomen.

4 **Melanostoma scalare.** A small, slim Hover-fly, which is very abundant and widespread. The larvae hatch from white eggs laid on the lower leaves of plants in damp country, and attack aphides and sometimes other insects. The flies can often be seen taking pollen from grasses.

5 **Cheilosia illustrata.** These flies, on the wing from April to September, resemble Bumble-bees (*Bombus sylvarum*) and can be seen taking nectar and pollen from the white umbels of flowers such as hogweed. The larvae feed on the roots of plants, and sometimes in the stems.

6 **Rhingia campestris.** Though ranging a considerable distance, this Hover-fly appears to breed only in cowdung in pastures. The white eggs are laid on clover and grasses overhanging cow-dung pats, into which the larvae drop and immediately burrow. They feed up rapidly, moving to the underside of the pat as it dries, and then rest for several days in the grass or clover roots before pupating. They emerge in numbers throughout the summer, but mainly in late spring and autumn. The adult has a projecting proboscis, which is normally folded but can be extended to suck long-tubed flowers such as bluebells. The fly often visits the large bindweed for pollen and will take shelter from rain in the trumpet.

7 **Xanthogramma pedissequum.** These striking, wasp-like flies are found from May to November in meadows and open woodlands from north Lancashire southwards. They visit flowers, usually flying low. The larvae have been found in ants' nests and probably feed on the aphides associated with ants (*see* p. 150).

8 **Volucella bombylans.** Both forms of this Hover-fly resemble Bumble-bees: the variety *bombylans*, with a red-haired tail end, resembles *Bombus terrestris* (p. 163), and the commoner variety *plumata*, with a yellow edge to its black abdomen, resembles *B. lapidarius*. The two forms often inter-breed. The larvae act as scavengers in the nests of *Bombus* or of the wasps, *Vespula germanica* (p. 154). The flies are found in open woodlands between May and August feeding at thistle, scabious, and bramble flowers.

9 **Volucella pellucens.** Another more common Hover-fly of open woodlands, whose larvae are scavengers in wasps' nests — in this case the wasp, *Vespula vulgaris*. The males hover conspicuously at a height of 8 or 10 feet, and visit a variety of flowers.

10 **Eristalis tenax** (Drone-fly). This fly, which is found in most parts of the world, resembles the Honey-bee (*Apis mellifera* p. 163). The white eggs are laid near manure heaps or on the edge of stagnant water, into which the larvae migrate. They swim freely, breathing through their long tails, which can be extended to 6 inches to reach the water surface — hence their name Rat-tailed Maggots. They move to drier earth to pupate. The adult males hover at 5 to 10 feet high in open spaces. Both sexes take nectar and pollen from flowers. The pollen grains are crushed between the toothed plates at the tip of the proboscis as the fly inflates and deflates this with blood. The pollen is partially dissolved during the process. Drone-flies can be seen throughout the year, for some overwinter in places such as caves and hollow trees.

LIFE SIZE (IN CIRCLES × 3)

1 Syrphus ribesii   3 Scaeva pyrastri   2 Syrphus balteatus

4 Melanostoma scalare   6 Rhingia campestris   5 Cheilosia illustrata

7 Xanthogramma pedissequum   8 Volucella bombylans

9 Volucella pellucens   10 Eristalis tenax

131

Nos. 1–5 belong to the family Syrphidae of Hover-flies, about which there is a note on p. 130. The Thick-headed Flies, family Conopidae (No. 6), have narrow waists, club-like abdomens, and antennae held directly forward, and somewhat resemble wasps. The Fruit or Gall Flies (Nos. 7–10) of the family Trypetidae are well-known because the larvae are plant pests.

1 **Sun-fly** (*Helophilus pendulus*). A very abundant fly, frequenting damp and marshy places and attracted to flowers. Its common name comes from its habit of not appearing until the sun is shining. The males hover over ditches, waiting for the females who, after mating, lay their eggs in such damp places. The larvae, like those of Drone Flies (p. 130), are 'rat-tailed' — that is, they have telescopic hinder ends. They can be found from May to November in rot-holes in trees as well as in stagnant waters. The adults fly from April to October.

2 **Large Narcissus Fly** (*Merodon equestris*). These flies appear in a variety of forms, which resemble various Bumble Bees. In spring the males fly low and hover over short, tufted grass and woodland tracks, awaiting the females. They appear to take possession of a territory, and if disturbed, return to it. The females lay their eggs near bulbs — wild bluebells or garden narcissus, especially varieties of *poeticus*. The larvae appear to enter the bulb through its soft neck or base, soon hollow it out, and live in the damp, decayed remains. They pupate in the soil either late in the autumn or the following spring. Adults are to be seen in greenhouses as early as February and fly out-of-doors from April to late August.

3 **Syritta pipiens.** A very common little Hover-fly, on the wing from March to October, when the sun shines, visiting various flowers. The females seek out wet, rotting vegetation, such as compost heaps, in which to lay their eggs, and in which the 'short-tailed' larvae live. The males hover round these heaps, making darts at other males and driving them away from their 'territory'. When a female appears, the male changes to a dancing flight, moving in an arc and flashing his silver face and patches on the thorax and abdomen in the sunshine, preparatory to mating.

4 **Xylota lenta.** This uncommon Hover-fly differs from the common *X. segnis* in having all black legs. They fly from May to September, visiting a few species of white flowers only. The males are usually seen sitting with closed wings on the leaves of shrubs and woodland plants. The larvae can be found in moist decayed wood.

5 **Criorrhina ranunculi.** A large Hover-fly which resembles those Bumble Bees which have similar yellow to red 'tails'. It appears in March, visiting the flowers of blackthorn and sallow and, later, hawthorn, in the sunshine. These Hover-flies generally keep to the top of trees and bushes, flying rapidly and dexterously, avoiding capture. The larvae live in wet, decaying matter in holes in trees.

6 **Conops quadrifasciata.** The wasp-like adults fly from July to October, probing the flowers of mint and thistles with their long probosci, and also visiting brambles, hogweed, and other flowers. The larvae are parasites on Bumble Bees. For some time the bee continues its foraging, while the maggot consumes its abdomen contents, at first breathing through its skin and then through its hind spiracles from the breathing system of the bee. Finally the bee collapses.

7 **Melieria omissa.** This little fly occurs in wet, marshy vegetation in May. Very little is known of the life history of the family Otitidae to which it belongs. The dagger-like ovipositor (egg-laying organ) of the females suggests that it may be used for inserting eggs in living vegetable matter.

8- **Fruit (Gall) Flies** (family Trypetidae). There are 74
10 species of these plant pests in Britain, some of which make galls. *Urophora cardui* (8) attacks the common creeping field thistle and so is harmless to man. The eggs are laid in the stem, causing a gall as large as a gooseberry, in which the larvae live. From one to seven grubs live in one gall and pupate there in April. The flies emerge about May, pushing their way out of the gall, which by then is hardened and brown. They make their escape by rhythmically inflating and deflating an air sac on their heads. They sit on the leaves of the thistle waving their wings in rhythm, and if disturbed, return to the same plant.

The larvae of the Hawthorn Fruit-fly (*Phagocarpus permundus*) (9) do not make galls but feed in hawthorn berries and in those of garden *Pyracantha* and *Contoneaster* in September and October. They pupate in the earth, and the flies emerge in June when they often enter houses and rest on window panes.

The larvae of the Celery Fly (*Philophylla heraclei*) (10) mine the leaves of plants of the parsley family, especially garden celery and parsnips and wild hogweed and angelica. Several maggots feed in one mine, causing the whole leaf to shrivel up. They pupate in the earth, and the adults usually emerge in June onwards. There are usually two broods, and there are two colour forms — reddish-brown or brownish-black.

LIFE SIZE (IN CIRCLES × 3)

1 Helophilus pendulus   2 Merodon equestris

4 Xylota lenta   3 Syritta pipiens   5 Criorrhina ranunculi

7 Melieria omissa   6 Conops quadrifasciata   8 Urophora cardui

9 Phagocarpus permundus   10 Philophylla heraclei

133

1 **Cheese Skipper** (*Piophila casei*). The larvae feed in cheese and move by clasping their hind ends with mouth-hooks and then suddenly releasing hold and springing forward — hence their name. They also feed on other stored food such as the scraps of ham and bacon about the slicing machines in shops. Other species of the genus frequent carrion in which the larvae feed.

2 **Carrot Fly** (*Psila rosae*). There are two forms of this member of the family Psilidae, one with yellow on the antennae and the other with all black antennae. The females of the latter can hardly be distinguished from *P. nigricornis*, the larvae of which also attack carrot-roots. The yellowish-white maggots can kill seedling carrots and stunt and cause forking in mature roots. They also attack hemlock, parsnip, and parsley. When fully-fed, they leave the carrots and pupate in the soil. There are two generations, the first adults emerging in late April, and the second at the end of July.

3 **Loxocera albiseta.** This member of the family Psilidae is recognized by its yellow face and scutellum. The female has a long ovipositor and resembles an ichneumon wasp (p. 147). The adults are on the wing from June to September. At present the life history is unknown, but the larvae probably live in plants, as do the larvae of the other two genera *Psila* and *Chyliza*, which live respectively in roots of plants or leaves of orchids.

4 **Snail Fly** (*Sepedon sphegus*). Members of the family Sciomyzidae have been called Marsh Flies; but they also occur on chalk downland away from marshes, and so are better termed Snail Flies because of the habits of the larvae, which attack snails. This species is a very distinctive fly to be seen southwards from Shropshire and Breconshire from the end of March until the end of October. The adults live in the marshy margins of ponds and lakes amongst the taller vegetation, often resting on the broad leaves of iris.

5 **Kelp (Seaweed) Fly** (*Coelopa frigida*). This is the largest species of the family Coelopidae, and the adults are to be found on wet, rotting wrack seaweed all through the year, but mostly in autumn and winter after large banks of wrack have been deposited during autumn gales. The eggs are laid and larvae feed in the wrack, helping in the decomposition of the seaweed, which otherwise tends to dry up. At times, there have been mass migrations inland, when the fly becomes a nuisance in shops, garages, and dry-cleaners; it is attracted by certain smells, especially that of trichlorethylene. In 1954 a notable invasion from the south coast brought them as far north as London.

6 **Killer Shore Fly** (*Ochthera mantis*). Some of the 120 or so British species of the family Ephydridae have larvae which feed on carrion or sewage; some mine in the meadow grasses; others live in the stems of water plants; but mostly the larvae are aquatic. This species captures small gnats and mosquitoes by means of spiny bristles on its legs, as it flies about the margins of acid heathland pools from August to October.

7 **Fruit and Vinegar Flies** (*Drosophila melanogaster*). The family Drosophilidae are well known for their attempts to share our food and drink, being attracted to the smell of beer, wines, vinegar, fruit juices, and fresh paint. *D. melanogaster* is especially fond of wine, and breeds quite prolifically on fermenting matter such as fruit pulp and yeast. Some members of the family, when in the larval stage, feed only on fungi.

8 **Holly Leaf Miner** (*Phytomyza illicis*). The larvae of this fly make yellowish blotches and lines, often edged with brown, on holly leaves throughout the country. There are over 300 species of this family, Agromyzidae, many of which confine themselves to one type of plant. The Holly Leaf Miner frequents the Holly during April to June, and the eggs are laid in the underside of the mid rib of the leaves, close to the base. The larvae feed in the leaves during September to May, even when the temperature is only just above freezing point. They pupate in the leaf-mine. This fly is abundant and widely distributed, even in the centre of towns, including London.

9 **Reed Gall Fly** (*Lipara lucens*). There are 132 British species of the family Chloropidae, which are particularly attracted to grasses, in which the larvae live. Some species cause much damage to cereal crops. The Reed Gall Fly is on the wing in June and July in areas where the reed *Phragmites communis* grows. The larvae form long, cigar-shaped galls on the stems, which shorten the stem and prevent flowering. The gall also acts as a home for the larvae of another Chloropid fly, *Crytonevra flavitaris*, but the latter remain in the outer layers of the sheath of the gall. The larvae pupate in the gall. Two other species of *Lipara* which are less common than *L. lucens*, also feed as larvae in the reed, but do not form galls.

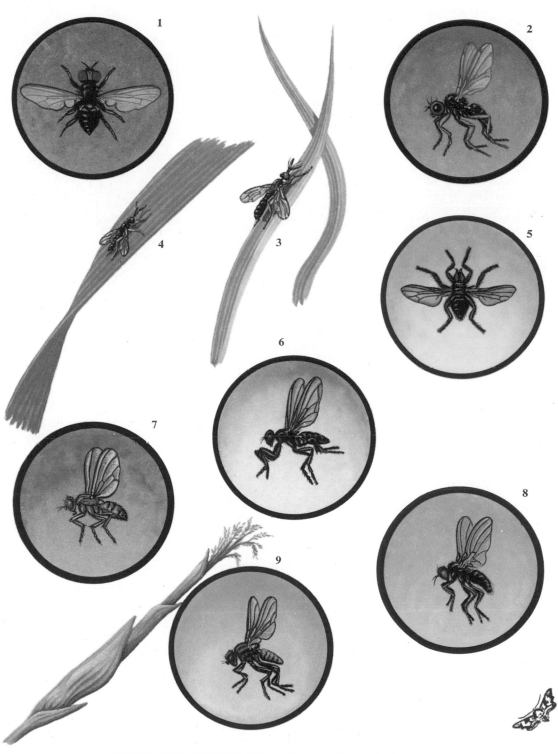

TWICE LIFE SIZE (IN CIRCLES × 4; Figs. 5, 6, 9 × 3)

1 Piophila casei   2 Psila rosae
4 Sepedon sphegus   3 Loxocera albiseta   5 Coelopa frigida
7 Drosophila melanogaster   6 Ochthera mantis   8 Phytomyza illicis
9 Lipara lucens

Nos. 3–10 on this page are Parasitic Flies of the family Tachinidae, of which there are 237 species in Britain. Most parasitize the larvae of butterflies or moths, but some attack those of beetles and several other kinds of insects. Most species lay their eggs on the skin of the host, but those of one genus, *Carcelia*, attach the eggs on stalks to the hairs of the host, and some species can pierce the host's skin and lay the egg under the skin. Others lay numbers of eggs on the host's foodplant, from which a proportion of the larvae reach the host.

1 **Bee Louse** (*Braula coeca*). This single member of the family Braulidae lacks wings and balancers, and resembles a louse (p. 19). The adult clings to a honey-bee, making its way to the bee's head, where it feeds on moisture from the bee's mouth-parts. The eggs are laid in the bee's hive inside the honey-cells, and the larvae feed on wax grains and pollen in the cell. They can pass between cells by tunnelling through the wax caps, and they pupate in the cells.

2 **Common Dung Fly** (*Scatophaga stercoraria*). This is the most common member of the family Scatophagidae. The bright yellow to ginger males and greenish-yellow females are familiar throughout the country, flocking to pats of dung in any field where cattle graze. They approach new deposits of dung from down wind, obviously scenting them, and they find tips of farmyard manure in the same way. Here they pair, sometimes two or three males trying to mate with one female. Eggs are laid on the dung, on which the larvae feed, pupating in the soil underneath. The adults also frequent flowers for nectar, and during dull periods will sit on herbage near the dung pats and attack other flies, cutting the nerve cord of their necks with the well-developed teeth near the base of their probosci, and then sucking their juices. They occasionally even kill other Dung Flies.

3 **Gonia divisa.** The female of this parasite of the larvae of the Turnip Moth, *Agrotis segetum* (p. 70), scatters large numbers of small eggs on turnips, docks, lettuce, or carrots, on which the Turnip Moth caterpillars feed. The adults are active from mid-March to early June in southern England as far north as Suffolk.

4 **Zenillia vulgaris.** A very common fly in woods and waste places throughout Britain from early May to mid-September. It is one of the commonest parasites on the larvae of many butterflies and moths and occasionally also of sawflies. The eggs are laid on the larvae and rapidly hatch. The first generation of larvae drop to the ground and pupate in the soil, but the second brood may hibernate in the host either as larva or pupa.

5 **Siphona geniculata.** A very common parasite of the crane-flies *Tipula maxima, T. oleracea* and *T. paludosa* (p. 122), and less commonly of caterpillars of the Small Tortoiseshell Butterfly (p. 48) and of moths of the sub-families Agrotinae and Larentiinae. The females lay their eggs on the larvae, and when these hatch they feed inside the host, taking air through the host's breathing system. The adults suck nectar with their long probosci from various flowers, especially hawk-weeds, ragworts, dandelions, buttercups, and wild parsnip, during the months of May to early November.

6 **Echinomyia fera.** A quite common parasite in marshes and open woodland as far north as Ross-shire, attacking the caterpillars of the Black Arches Tussock and other moths. The larvae and pupae are attached to the entrance hole in the host to enable them to breathe. The adults fly between early April and mid-September, and are to be seen most commonly in late summer visiting, among other flowers, water mint and angelica in marshes and devil's-bit scabious on wood margins.

7 **Echinomyia grossa.** This is the largest British parasitic fly, and it flies and looks like a bumble bee. It is found chiefly about sandy heaths and woodlands from June to September as far north as Ross-shire. It is a parasite of the caterpillars of larger moths such as the Pine Hawk and Oak Eggar (pp. 60, 90). The larvae attach themselves to the host as it feeds, and burrow just inside the skin, breathing through the entrance hole. The adults take nectar from flowers such as sea holly on the coast and ragwort, angelica, wild parsnip, and marsh thistle.

8 **Eriothrix rufomaculatus.** Only the sub-species *monochaeta* occurs in Britain, and this is common as far north as Inverness-shire from June to September. The adults fly briskly over and through flowers and herbage, visiting flowers such as ragwort, creeping thistle, wild parsnip, and hogweed. This parasitic fly has never been known to breed in Britain but on the continent it may breed on the larva of a Tiger moth, *Arctia hebe*.

9 **Dexiosoma caninum.** The males of this long-legged fly are commonly found resting on the top fronds of bracken in open woodland from June to September. It parasitizes the Cockchafer beetle larvae (p. 182) in England and Scotland, but its method of attack is unknown.

10 **Alophora hemiptera.** The dark-winged form, when the wings are closed, looks rather like a plant-bug (p. 27), and is thought to parasitize certain bugs, grasshoppers, and crickets. In France it is known to parasitize the plant-bug *Palomena prasina*. The adults vary in size and wing colour. The female has a piercer-type ovipositor. *A. hemiptera* is uncommon, being found locally in scrub and marginal woodland, especially marshy areas, as far north as Yorkshire, from April to August. It has been seen on the flower heads of angelica and hogweed.

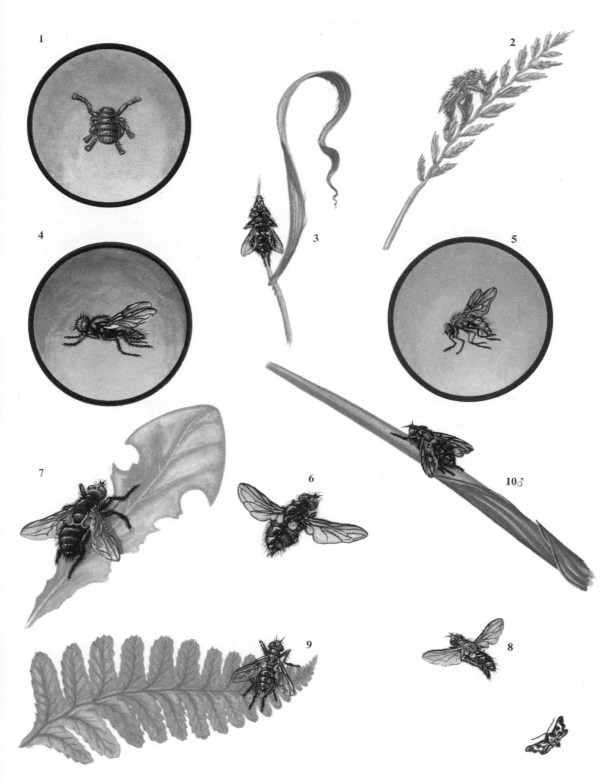

LIFE SIZE (IN CIRCLES × 3; *Fig.* 1 × 6)

1 Braula coeca   2 Scatophaga stercoraria

4 Zenillia vulgaris   3 Gonia divisa   5 Siphona geniculata

7 Echinomyia grossa   6 Echinomyia fera   10 Alophora hemiptera

9 Dexiosoma caninum   8 Eriothrix rufomaculatus

137

BLOW FLIES,
WARBLE FLIES, ETC.

The 100 species of the family Calliphoridae (1–5) include some of the best known flies, pests in houses. Most are carrion-feeders, some laying their eggs on meat. They vary in habits from scavengers to parasites. The families Tachinidae (6–7) and Muscidae (8) include members that attack animals such as cattle, sheep, and horses.

1 **Flesh Fly** (*Sarcophaga carnaria*). A very common fly from April to October as far north as Morayshire. The maggots of this and others of the large genus feed on carrion, ranging from snails and birds to unprotected joints of meat. The eggs hatch in the female's body, and the maggots are deposited on the food. They emit a liquid in which they live, being protected from drowning in it by specially adapted breathing holes. They pupate in the soil, and the flies emerge by constantly inflating and deflating air sacs on their heads and so forcing a way through the soil.

2 **Blow Fly or Bluebottle** (*Calliphora erythrocephala*). A very abundant house pest. There are two species confined to northern Scotland, and a third, *C. vomitoria*, with red-haired black cheeks, which is very common as far north as Sutherland. They are about at all times of year, females often to be seen indoors in the winter or outside on warm days. The females lay large numbers of eggs and are often searching urgently for suitable places to lay them. They will even inject them through a mesh cover on to cooked meat. Both eggs and larvae need damp conditions, and so the larvae often travel a long distance to find drier conditions for pupating.

3 **Phormia terrae-novae.** This shiny blue fly is found commonly throughout Britain, though not in such abundance as Bluebottles. The larvae live in the nests of birds such as sparrows and swallows, sometimes in hundreds, and they attack the newly hatched nestlings, sucking their blood and sometimes killing them.

4 **Greenbottle** (*Lucilia caesar*). This very common fly is found throughout the country in gardens, wayside, and woodland from April to October. The female lays her eggs on carcases, seeking the soft parts that are shaded from the sun. The maggots feed in the same way as Flesh Flies (1) do, and the adults visit a large variety of flowers for nectar. The related *L. sericata* often lays eggs on the sores of sheep, and the larvae live in the open wounds, sucking blood.

5 **Cluster-fly** (*Pollenia rudis*). A very common parasite of earthworms. Eggs are laid in the earth, and the larvae enter the earthworms, one per worm, living there for several months until they have eaten almost the whole contents of the worm. Cluster-flies are distributed throughout Britain and have been seen on the wing in every month, though they are most abundant in spring

and late summer. The males congregate in small parties in the sunshine, individuals rising from time to time to execute a figure of eight and, as they return, flashing the golden-green colour of the thorax. This may be some kind of courtship display flight. In the autumn, large numbers gather on the warm sides and roofs of houses, from whence they crawl into attics and such places. This is how they have gained their common name of Cluster-fly. Like *Musca autumnalis* (p. 140), they also shelter in crevices in the bark of trees.

6 **Ox Warble Fly** (*Hypoderma bovis*). A fly found in June and July from Surrey to as far north as the Forth basin. The eggs are attached to hairs on the legs of cattle, and occasionally on horses and goats. The larvae pierce the skin at the base of a hair and move as they feed until they reach the back of the animal. Here they rest, soon causing a small swelling — the warble — with an opening to the air, in which they live. When fully fed, they drop to the ground and pupate. The related *H. lineatum* is less restricted to cattle, and a third British species, *H. diana*, is found in Scotland attacking roe and red deer.

7 **Sheep Nostril Fly** (*Oestrus ovis*). This pest of domestic animals is found from Kent to the Forth area in Scotland, from May to September. It suns itself on walls, fences, and stones near sheep pastures. The eggs hatch in the female's body, and the larvae are deposited singly by her when in flight, near the sheep's nostrils. They then crawl into the nasal passages, pierce the lining, and suck the blood until mid-winter. Some then move to the front of the sheep's skull. All return when fully developed to the nostril, drop to the ground, and pupate from April to June. A single female may deposit as many as 500 larvae.

8 **Horse-bot Fly** (*Gasterophilus intestinalis*). This belongs to a separate sub-family in Muscidae — the Gasterophilinae. The flies have a habit of gathering about the tops of mountains. The females have long ovipositors and, when laying their eggs, attach them to the hairs on horses. They also have been found on mules and donkeys, especially during the First World War, when there were many of these in use. The eggs are placed about the shoulders, belly, and upper hind legs and get rubbed by the horse's lips. The larvae cling to the lips, and in that way enter the body. They move to the stomach, hook on to the lining, and feed until ready to pupate, when they pass out in the dung of the horse in July and August, and pupate in the dung.

LIFE SIZE

1 Sarcophaga carnaria   3 Phormia terrae-novae   2 Calliphora erythrocephala

4 Lucilia caesar   6 Hypoderma bovis   5 Pollenia rudis

7 Oestrus ovis   8 Gasterophilus intestinalis

HOUSE FLIES, FLOWER FLIES,
AND BITING FLIES

House Flies and other members of the family Muscidae (Nos. 1–5) are commonly seen about houses and farms. The larvae mainly feed on excrement, carrion, and decaying vegetable matter, and the adults feed in the same places and, though they do not bite, spread disease and should be destroyed whenever possible. The closely-related Flower Flies (Nos. 6–8) of the family Anthomyinae are vegetable eaters as larvae, some being pests to farmers and gardeners. The biting flies of the families Hippoboscidae (Nos. 9–11) and Nycteribiidae (No. 12) attack birds and mammals. The eggs hatch in the female's body, and the larvae, when dropped, pupate almost immediately.

1-2 Musca domestica and autumnalis. Both these are very common throughout Britain. The House Fly (1) is a serious menace to health, carrying disease from the filth in which it breeds to our food. The female lays batches of 100 – 150 eggs which, in warm, damp conditions may complete development and produce new flies in a week or so. In colder conditions the process is slower. As a single female may lay 900 eggs, the rate of increase can be terrific.

The Autumn Fly (2) is on the wing from February to October. The eggs are laid in cow pats, in which the larvae feed. The adults collect in the sunshine on tree trunks and warm walls and, at dusk, take shelter in the cracks. In autumn they enter houses to hibernate.

3 Stable Fly (*Stomoxys calcitrans*). This differs from a House fly in having a stiff, pointed proboscis. The larvae feed in stable litter but not in cattle dung. They pupate in the ground or in dry litter. The adults feed by piercing the underparts of horses, cattle, and sometimes dogs, cats, and chickens, from May to October, and may attack man, usually outdoors.

4 Sweat Fly (*Hydrotaea irritans*). A very common fly throughout Britain from May to October. On a hot day the females buz in clouds round the heads of people walking through woodland, not biting, but sipping perspiration on their necks. They bother horses and cattle, on the dung of which the larvae feed.

5 Lesser House Fly (*Fannia canicularis*). Common visitors to houses, these flies are abundant in suitable places all through the year. The larvae feed in dry rubbish, but the flies do not usually carry disease as House Flies do. The males are generally to be seen circling, often under trees.

6 Wheat Bulb Fly (*Leptohylemyia coarctata*). A serious pest of wheat, and also rye and barley, throughout Britain. The females lay their eggs from late July to August on bare soil. They hatch in very early spring, and the maggots burrow into the centre of new shoots, where they feed until they pupate in the soil in the late spring. The adults emerge in June and may be found on the tops of the crop in calm weather, but move close to the ground when the wind rises.

7 Cabbage Root Fly (*Erioischia brassicae*). The adults start laying eggs from mid-April to May on the bare soil about the roots of cabbage crops. They soon hatch, and the larvae feed in the roots and in cracks in the main stem. There appear to be at least three overlapping generations, and larvae may still be feeding until January; adults may be about the fields from April until early October.

8 Onion Fly (*Delia cepetorum*). A common and widely distributed pest to all growers of onions and shallots. The adults emerge in May and June, feed at flowers, and within a fortnight the first eggs are laid. These are placed on the young leaves or neck of the onion, and hatch after about three days. The larvae move down and enter the bulb through the roots, causing the leaves to wilt and die and the bulb to soften. There are three overlapping generations, and the pupae of the third generation hibernate in the soil.

9 Forest Fly (*Hippobosca equina*). A rare fly except in the New Forest and parts of Dorset, Hampshire, and Wales. It attaches itself to horses, especially New Forest ponies, donkeys, cattle, and dogs, and also to humans passing through bracken; it causes irritation to animals generally between the thighs or under the tail. The females drop living larvae on decaying humus under bracken.

10 Bird Parasite (*Ornithomya avicularia*). This fly has been found in Britain southwards from Cumberland on 43 species of birds, but chiefly on rooks, jackdaws, and starlings. The flies are generally noticed when they leave dead birds as they grow cold, or on the arms of people placing rings on birds. The females drop full-grown larvae which pupate almost at once, often in birds' nests, where the flies attack young birds.

11 Sheep Ked (*Melophagus ovinus*). This fly, like the Bee Louse (p. 136), lacks both wings and balancers (*halteres*). It is found in any month in the wool of sheep. The females may live as long as 4 months and produce up to twelve larvae, which are laid singly and pupate immediately in the wool.

12 Bat Fly (*Nycteribia biarticulata*). Looking like a tiny six-legged spider, this fly feeds on horse-shoe bats by piercing the skin and sucking blood. The females produce living larvae which they attach to beams or walls in the bat roosts, and which pupate at once.

LIFE SIZE (IN CIRCLES × 3)

1 Musca domestica   2 Musca autumnalis
4 Hydrotaea irritans   3 Stomoxys calcitrans   5 Fannia canicularis
7 Erioischia brassicae   6 Leptohylemyia coarctata   8 Delia cepetorum
10 Ornithomya avicularia   9 Hippobosca equina
12 Nycteribia biarticulata   11 Melophagus ovinus

The sub-order Symphyta, of which there are nearly 500 British species, can be distinguished from the Apocrita (wasps, bees, and ants) by having no narrow 'waist' between thorax and abdomen. Also the larvae feed on plants: either on leaves or by boring in buds, flowers, fruit, stems, or even wood. The wood-borers are known as 'Wood-wasps' (Nos. 1–3), and all the rest as Sawflies (Nos. 4–9). The name Sawfly was given because the female ovipositor in many species has developed two saw-like structures with which she embeds her eggs in plant tissues. The larvae of many species are both more conspicuous and more harmful than the adults. They often look much like the larvae of butterflies and moths. The adults are short-lived, the males dying a few hours after mating.

1　Xyela julii. This species represents what is thought to be the most archaic family of the order Hymenoptera. The adults, less than 5 mm long, are most commonly found feeding on pollen at birch and other catkins, sometimes in great abundance. The virtually legless larvae feed in the developing male catkins of pine before its pollen is shed.

2　Pamphilius sylvaticus. This Wood-wasp, with its flattened body and broad wings, represents another primitive family, which appears early in the summer and flies very fast in the sunshine. The larvae have long legs on the thorax but only vestiges of legs on the abdomen. They live on trees of the Rose family in solitary leaf-rolls, which they construct and hold together with silk.

3　Giant Wood-wasp (*Urocerus gigas*). A member of one of the two British families of wood-wasps and can easily be mistaken for a social wasp (*see* p. 155). What is thought to be the female's sting is really her ovipositor with which she lays her eggs deep in the wood of sickly or recently felled coniferous trees. This has led to the common name Horntail. The larvae may take up to three years to mature, during which time the felled trees in which they live may have been sawn into planks and used for building houses or making packing cases, or may have been exported. One species of Wood-wasp was recently introduced accidentally into pine plantations in New Zealand and Australia, where periodic droughts make the trees specially susceptible to insect attack, and there it caused great havoc.

4　Wheat-stem Borer (*Cephus pygmaeus*). A well-known pest of corn and forage grasses, belonging to a family whose larvae all bore in stems of grasses and other flowering plants and shrubs. They have narrow cylindrical bodies and long antennae.

5　Rose Sawfly (*Arge ochropus*). This is a well-known pest of garden roses. The eggs are embedded, fifteen to twenty together, in a double row in the stem of a growing rose shoot. The larvae, which feed on the leaves in a conspicuous party, are green and yellow flecked with black, bristly warts. These Sawflies represent a large family, widely distributed in the warmer parts of the world; the adults of all the species have only three-segmented antennae.

6　Abia sericea. This beautiful Sawfly and the three following species belong to the family Cimbicidae, which is characterized by short, club-shaped antennae. This one is to be found in hedgerows and permanent grassland, and the larvae feed on scabious.

7　Large Birch Sawfly (*Cimbex femorata*). This is one of the largest of British Sawflies and is attached to birch trees. Its larvae, which feed solitarily on the leaves in late summer, are pale grey-green as though dusted with white powder and have a dark mid-dorsal line. The adults of this genus are easily distinguished from those of Trichiosoma because they have large pale membranous gaps between the first and second abdominal segments and almost hairless, shining abdomens.

8　Large Hawthorn Sawfly (*Trichiosoma tibiale*). This and *T. lucorum* are among the commonest of the large Sawflies. The larvae feed on hawthorn, and the large, oblong parchment-like cocoons (8a) can often be found attached to twigs in winter hedges. The larvae are bright greenish yellow, as though dusted with white powder, and can easily be distinguished from those of *T. lucorum* because they lack the dark line along the back. The adults are clothed in soft down, and also there are close, short, dark hairs on the abdomen. They do not have a pale gap between first and second segment of the abdomen, which *T. lucorum* has.

9　Trichiosoma lucorum. This Sawfly is probably not really a distinct species from *T. tibiale*. It does not have, however, the darkened tibiae of the hind legs, and it does have pale hairs on the abdomen. It is found attached to sallow and birch.

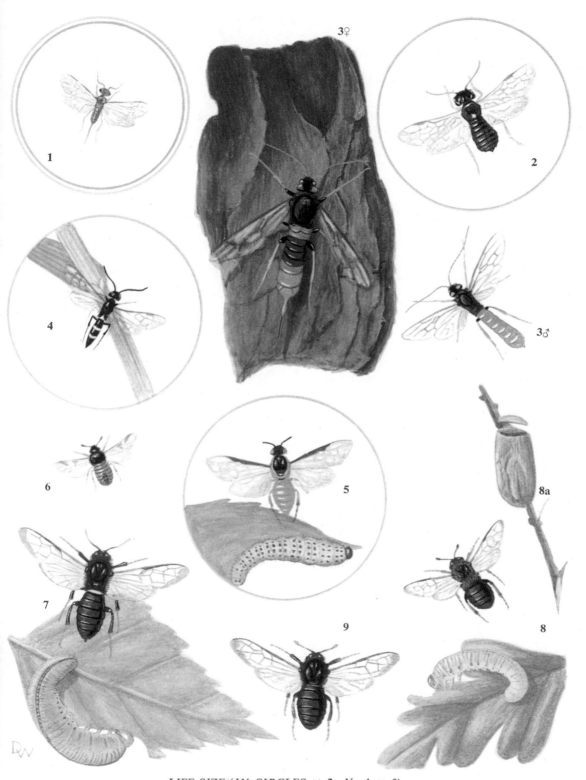

LIFE SIZE (IN CIRCLES × 2; *No.* 1 × 3)

1 Xyela julii   2 Pamphilius sylvaticus

4 Cephus pygmaeus   3 Urocerus gigas

6 Abia sericea   5 Arge ochropus + larva

7 Cimbex femorata + larva   9 Trichiosoma lucorum   8 Trichiosoma tibiale + larva

For a general note on Sawflies see p. 142.

1 **Pine Sawfly** (*Diprion pini*). This species belongs to a small family in which the larvae feed entirely on the needles of coniferous trees. Some species of this family are serious forestry pests in northern temperate forests and plantations, and certain European species accidentally introduced into Canada have caused great havoc there. They are characterized by their many-segmented antennae, serrated in the female and comb-like in the male.

2 **Allantus cinctus.** This and the rest of the species on this page belong to the family Tenthredinidae, the largest Sawfly family with more members than all the other families put together. The larvae of *A. cinctus* feed on the edges of rose-leaves and usually overwinter in tunnels bored into the pith of twigs exposed by pruning. Like most members of the family, the adults are sluggish fliers, not usually flying more than a few yards. On hot days they may be seen resting on flowers.

3 **Heterarthrus aceris.** This species represents the type of Sawfly the larvae of which live in blister leaf-mines, between the upper and lower epidermis (outer skin) of leaves. The larva of *H. aceris*, when fully fed in its mine in a sycamore leaf, spins a circular 'hammock' round itself, attached to the upper epidermis of the leaf. The whole disc (Fig. 3a) thus formed then drops from the leaf, and the larva inside, by alternately relaxing and tensing its body, is able to make the disc jump, and so move to a suitable resting place for overwintering.

4 **Poplar Sawfly** (*Trichiocampus viminalis*). The yellow, black-flecked, bristly larvae of this species are a familiar sight feeding in midsummer in parallel rows from the under surfaces of poplar leaves; they leave transparent 'windows' of the upper epidermis in the leaf intact. When fully fed, the larvae move on to the bark of the tree, where they form parchment-like cocoons massed together against the bark.

5 **Apple Sawfly** (*Hoplocampa testudinea*). This is an example of a group of Sawflies which lay eggs in the ovaries of flowers of the rose family. Their larvae feed in the developing fruit. The Apple Sawfly is a well-known apple pest, causing infested apples to fall from the trees at the end of June, and those less damaged apples that remain on the tree have superficial scars that ruin their market value.

6 **Pontania vesicator.** The species of this genus make galls for their larvae on the leaves of willows and sallows. These are the only Sawflies which make leaf-galls, though there are others whose larvae live in swollen buds or stems. This species makes broad-bean-shaped galls in the leaves of the purple willow in northern Britain. Other species make pea-shaped or coffee-bean-shaped galls, each Sawfly species being restricted to a separate species or group of willows.

7 **Palisade Sawfly** (*Stauronematus compressicornis*). The green larvae of this inconspicuous black Sawfly feed on the edges of poplar leaves. They have the interesting habit of raising a palisade of dried saliva spots round the part of the leaf on which they are feeding. The palisade looks like a growth of mould and is thought to deter hungry marauders that hunt by sight or touch, by suggesting that the larvae are not quite fresh or wholesome enough to be eaten.

8 **Currant and Gooseberry Sawfly** (*Nematus ribesii*). This is the commonest of four most frequent Sawfly pests of currant and gooseberry bushes. The larvae feed quickly in massed groups through three or four broods a year. In warm summers their numbers increase enormously, and currant and gooseberry bushes become stripped of their leaves.

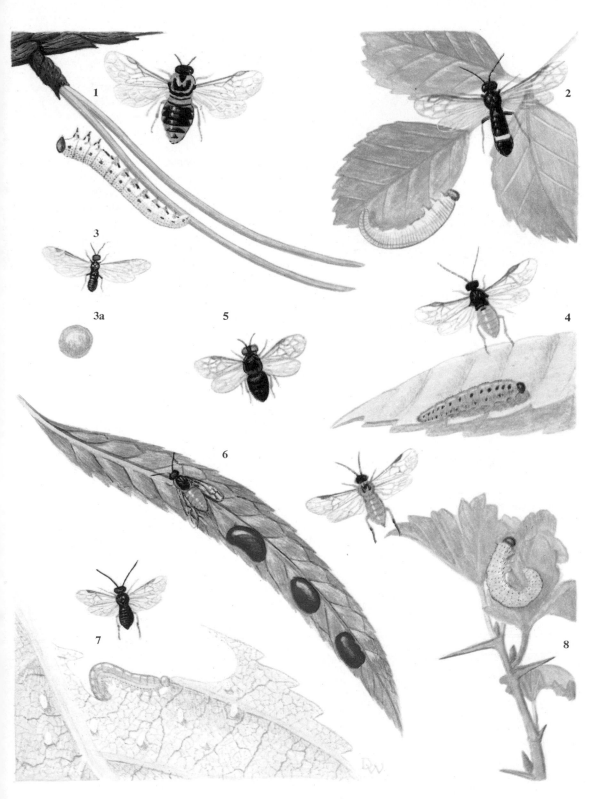

TWICE LIFE SIZE

1 Diprion pini + larva 2 Allantus cinctus + larva

3 Heterarthrus aceris 5 Hoplocampa testudinea 4 Trichiocampus viminalis + larva 145

7 Stauronematus compressicornis + larva 6 Pontania vesicator 8 Nematus ribesii + larva

BRACONIDS, ICHNEUMONS, AND CHALCIDS

All these three groups consist mostly of species which are parasites, that is, they lay their eggs in or on other insects (usually the larvae), spiders, or mites, on which their larvae feed. Some are useful to man in reducing the numbers of harmful insects. Some, however, lay their eggs in plant tissue, on which the larvae feed. Some are hyperparasites — that is, parasites on other parasites that have already attacked the host. They vary greatly in size and in way of life.

The families Braconidae and Ichneumonidae are closely related, but differ in the pattern of wing veins. The Braconids (Nos. 1–2) include both solitary and gregarious species. Most Ichneumons (Nos. 3–9) attack the larvae of butterflies and moths, though some attack spiders. Most are solitary, but a few develop gregariously in the host's cocoon before and after pupation.

Chalcid Flies (Nos. 10–11) are mostly very small and metallic blue or green in colour, with very reduced wing veining. There are an enormous number of species, most of them parasites on insects, and of great value in controlling the insect pests which attack our food supplies.

1 **Apanteles glomeratus**. A little black Braconid which can be seen in summer among cabbage patches seeking out the caterpillars of the Large White Butterfly (p. 55), in which it lays its eggs. Later, its sulphur-yellow cocoons are to be seen in masses on adjacent walls and fences.

2 **Protomicropolitus connexus**. This species is parasitic on moths of the family Lymantriidae (p. 68), and is deliberately used by man in order to control the numbers of these moths. The larvae spin their cocoons in a dense mass within those of the host.

3 **Amblyteles armatorius**. This is one of the larger Ichneumons, which lays its eggs on the chrysalids of the moth family Noctuidae (*see* pp. 70 – 88). It is more than half-an-inch long and black, with a yellow spot on the rear part of the thorax, and with two segments of the abdomen yellow. It has long antennae and black and yellow legs. Superficially it resembles a pompilid wasp (*see* p. 154), for the ovipositor is not conspicuous as it is in most Ichneumons.

4 **Amblyteles elongatus**. This is about the same size as *A. armatorius*, but is entirely black, except for a small yellow scale on the back and yellowish-red legs. It also is a parasite on Noctuid moths.

5 **Gelis melanocephala**. The females of this tiny, ant-like species are wingless, and they attack the eggs of spiders in their cocoons. Some of the spiders they attack form their cocoons beneath stones, and the ant-like shape of the parasite enables it to creep into such underground situations.

6 **Apechthis rufata**. This black, red-legged Ichneumon is about half-an-inch long, and has a very stout ovipositor which is so jointed that it can be inserted into the host at right-angles to its body. A favourite host is the Large White Butterfly larva (p. 55). The Ichneumons pupate in small golden-yellow cases massed together on palings or walls, with often the skin of a caterpillar among them.

7 **Rhyssa persuasoria**. A large Ichneumon, well over an inch long, and with an ovipositor twice as long as that. Its body is black with the margins of the abdominal segments pale yellow. It is a parasite on the Giant Wood-wasp, *Urocerus gigas* (p. 143). The Wood-wasp larva burrows in the timber of pine trees; yet the Ichneumon succeeds in locating it, and then drills through the wood to reach its victim.

8 **Lampronata setosas**. Another large Ichneumon, parasitic on the Goat Moth (p. 113). It is about three-quarters-of-an-inch long, the antennae and ovipositor adding a further inch. The wings are cloudy and the legs reddish. The long ovipositor, like that of *Rhyssa*, is necessary in order to reach the host larvae, which live in wood.

9 **Ophion luteus**. One of the large Ichneumons which has probably been seen by everybody at some time owing to its habit of flying into houses, especially at dusk. It is yellowish brown all over and is about three-quarters-of-an-inch in length. The female has no long ovipositor as most Ichneumon females have. It is a parasite on caterpillars of several Noctuid moths, including the Broom Moth (p. 75) and the Sycamore Moth (p. 81), and it also attacks Puss Moth caterpillars (p. 65).

10 **Prestwichia aquatica**. This tiny Chalcid is a parasite of water beetles (p. 169) and bugs (p. 31), attacking their eggs. It belongs to the family Trichogrammatidae, which includes some of the smallest examples of insects. It has very distinctive wings, which are extremely slender and fringed all round with long hairs; these enable the insect to swim under-water in order to reach the host eggs.

11 **Pteromalus puparum**. A little Chalcid which lays its eggs in the soft-skinned, newly-formed chrysalids of Cabbage White Butterflies (p. 55). One egg per caterpillar is sufficient because each egg divides to give rise to many larvae; therefore, in the following spring, fifty to a hundred adult Chalcids may emerge from a single caterpillar host.

LIFE SIZE (IN CIRCLES × 6; No. 10 × 12) — All females

1 APANTELES GLOMERATUS	3 AMBLYTELES ARMATORIUS	2 PROTOMICROPOLITUS CONNEXUS
4 AMBLYTELES ELONGATUS	5 GELIS MELANOCEPHALA	6 APECHTHIS RUFATA
8 LAMPRONATA SETOSA	7 RHYSSA PERSUASORIA	9 OPHION LUTEUS + PARASITIZED LARVA
10 PRESTWICHIA AQUATICA	11 PTEROMALUS PUPARUM	

Nos. 1–6 on this page are the gall-making group of the super-family Cynipoidea, the majority of which are parasites on other insects. This group either make galls on trees or plants, mainly oak or rose, or they live in the galls of other species. They are mostly small or very small, and are usually black, red, or yellow, or a mixture of these colours, and are very difficult to tell apart without a good microscope. They are, in fact, easier to identify from their galls which vary considerably in size, shape, colour, and position on the host plant.

Some gall-makers have alternate non-sexual and sexual generations. The non-sexual (or agamic) generation is without males. It is only recently that these alternate generations have been realized to be the same species: they used to be thought of as different species or even different genera. The females lay their eggs in the plant tissues, and the activity of the grub in the growing plant tissue forms the galls, though exactly what causes each gall-maker to form its own particular type of gall is not yet fully understood. We use the symbol ⚥ to denote the non-sexual generation and the symbols ♂ ♀ for the sexual generation.

Nos. 7–9 are parasitic stinging ants and wasps, in which the basal segment of the abdomen is nearly always constricted to form a waist.

1 **Robin's Pin-cushion Gall** (*Diplolepis rosae*). This is found on twigs and leaves of roses, especially wild roses. It is closely covered with very long, branched, reddish hairs and is brightly coloured. The galls mature in May, when the adults emerge. This and the next species have only sexual generations, but there are many more females than males.

2 **Spiked Pea Gall** (*Diplolepis nervosa*). The galls are like peas with spikes, though these are occasionally absent; they are found on the leaves of roses. They mature in May.

3 **Oak-apple Gall** (*Biorhiza pallida*). These irregular, globular-shaped galls (3a) mature in May to June and become brownish-yellow with age. The adults which emerge from them are winged males and wingless females. These lay eggs in the oak roots, making root galls (3b). From these come a generation of ⚥ females, which climb the tree and lay their eggs in the oak buds to produce oak-apple galls again.

4 **Marble Gall** (*Andricus kollari*). The spherical, single-celled, waxy galls, though pale green when young, become reddish-brown when mature, and grow in the base of developing buds. They mature in September and produce ⚥ females. The galls which come from their progeny mature in May or June, and from them emerge both males and females.

5 **Spangle Galls** (*Neuroterus quercusbaccarum*). These very common Gall Wasps lay their eggs in the veins of oak leaves, and lentil-shaped galls (5a, enlarged and natural size) appear in August, becoming reddish when mature. The grubs overwinter in the galls, pupate there, and emerge in April as ⚥ females. These lay eggs deep in the catkins, and the grubs form spherical 'currant galls' (5b) on the catkins and leaves in May and June. From these emerge the generation of males and females — though mainly females.

6 **Silk-button Gall** (*Neuroterus numismalis*). These Gall Wasps of oak trees follow the same pattern of life as those of the Spangle Galls. The hemispherical hairy galls with central pits (6a) mature in September, and the Blister Galls (6b) mature in May and June.

7 **Ruby-tail Wasp** (*Chrysis ignita*). This is the commonest of a group of about 30 brilliantly-coloured British species of Cuckoo-wasps, which are parasitic in the nests of various bees and wasps. Other species may be entirely blue or green, or have different parts of the body red, blue, or green, and some are much smaller than *C. ignita*. They are to be seen only in bright sunshine and are extremely active.

8 **Methoca ichneumonides.** A rather primitive wasp of which the black, Ichneumon-like male has a long upwardly curved spine, and the much smaller, wingless, red-and-black, ant-like female has a red thorax quite as long as its abdomen and superficially divided into three leg-bearing parts. It lives in sandy heathland, where the females parasitize the larvae of Tiger-beetles (p. 165). It allows itself to be caught by the ferocious larva and dragged into the burrow. There, it paralyzes the larva, lays an egg on its body, escapes, and seals up the burrow with a stone which it has already provided.

9 **Mutilla europaea.** The long red thorax of this striking-looking wasp does not show the usual divisions, and only the male has wings. The deep-blue abdomen has a band of thick, pale hairs across the first segment, and interrupted bands across the second and third segments. This species is a parasite in the nests of Bumble Bees (p. 163).

TWICE LIFE SIZE (IN CIRCLES × 3); GALLS LIFE SIZE

1 Diplolepis rosae   2 Diplolepis nervosa   3 Biorhiza pallida
5 Neuroterus quercusbaccarum   6 Neuroterus numismalis   4 Andricus kollari   149
8 Methoca ichneumonides   7 Chrysis ignita   9 Mutilla europaea

All ants have large heads and the basal segments of the abdomen constricted to form one or two 'waists'. There are 41 British species, most of which are social insects, living a highly-developed community life. An ant colony consists of perfect females or queens (♀) which lay eggs; males (♂), which die after fertilizing the females; and imperfect females or workers (☿), which do all the work. The males and females have wings, but after the marriage flight, the queens remove theirs and remain in the nest. The workers, the bulk of the population and wingless, have specialized duties: some forage for food, which they carry to the nest in their 'crops' and pass on from mouth to mouth; some care for the eggs and larvae, moving them about the nest to obtain the correct amount of warmth, moisture, and air; some attend to the queens; some build, repair, or extend the nest; and others guard the colony. Both queens and workers of many species have stings or a means of squirting formic acid at an enemy from their tail ends.

The nest is simple compared to a bee's or wasp's nest, being usually a series of galleries and chambers excavated in earth or wood. As ants do not build cells for the grubs, if the nest is disturbed, they can fairly easily carry the eggs, larvae, and pupae to a new place. If not disturbed, the nest can carry on almost indefinitely, new queens being produced at intervals. In the laying period a queen lays eggs at the rate of one every 10 minutes, and she may continue for 6 or 7 years. When the colony grows too big, queens with groups of workers move off to start fresh colonies.

1 **Anergates atratulus.** Curious and somewhat repulsive ants living only in the nests of the ant, *Tetramorium caespitum* (No. 2). They produce no workers of their own, and the males are wingless, pupa-like creatures, scarcely able to walk. For their well-being they rely entirely on the *Tetramorium* workers, which kill their own queen once their nest has been invaded. With no queen, the host colony gradually dies out, but not before rearing the next generation of *Anergates* parasites, whose queens become hugely distended with eggs. These ants are rare and occur only in Dorset and south Hampshire.

2 **Tetramorium caespitum.** A sturdy little brown-black ant, commonly found nesting under turf and stones on sandy heathlands in southern England. The shining black queen and male are much larger than the workers, the queen twice as large. Adjacent colonies are constantly at war with each other, for the workers are fierce and quarrelsome. Despite this, they give hospitality to the tiny ants, *Strongylognathus* (No. 4), as well as to the deadly *Anergates atratulus* (No. 1)

3 **Pharaoh's Ant** (*Monomorium pharaonis*). This little yellow species originally came from hotter climates and was accidentally introduced into Britain about 150 years ago. Since then it has spread, and is to be found in most towns and cities, being able to survive only in artificially-heated premises.

4 **Strongylognathus testaceus.** A tiny yellow ant to be found only in the nests of *Tetramorium caespitum* in Dorset and south Hampshire. It produces its own workers, and its colonies seem to live in harmony with their hosts, being distinguished by size and colour.

5 **Leptothorax acervorum.** The largest of four small, reddish-yellow ants, all of which have smooth, shining

abdomens but hairy and quite coarsely sculptured head and thorax. They nest under bark and stones.

6 **Red Ant** (*Myrmica ruginodis*). One of the more common ants of the countryside which nests beneath stones, in rotting stumps, or sometimes sharing an earth mound with *Lasius flavus* (No. 8). It has long, sharp spines on the last segment of the thorax. It is ants of this genus which take the larvae of the Large Blue Butterfly into their nests to 'milk' them (*see* p. 52).

7 **Lasius fuliginosus.** A shining black ant with a large, heart-shaped head, which is not common and is not found in Scotland and much of Ireland. In places where there are large colonies, often several nests close together, long files of workers can be seen ascending every tree in the neighbourhood searching for aphids, the excreta of which seems to be their principal food. The nests, usually in hollow trees or among tree roots, have galleries made of chewed wood fragments which dry hard. The queens cannot start their own colonies, but start in nests of relatives.

8 **Meadow Ant** (*Lasius flavus*). This ant spends most of its life below ground, and is responsible for the grass-covered mounds so often seen in uncultivated fields and orchards. Unlike *Myrmica* (No. 6), which lives in similar situations, it has only a single abdominal 'waist'. It 'farms' an underground aphid, which it milks for a sweet liquid it secretes (*see* p. 34).

9 **Garden Black Ant** (*Lasius niger*). This is the ant which invades kitchens and larders and whose marriage flights are sometimes seen in such numbers that they disrupt traffic in busy cities, especially when these flights coincide with those of *L. flavus*, as they often do. Their nests are beneath stones or logs, or under paving stones or the masonry of buildings.

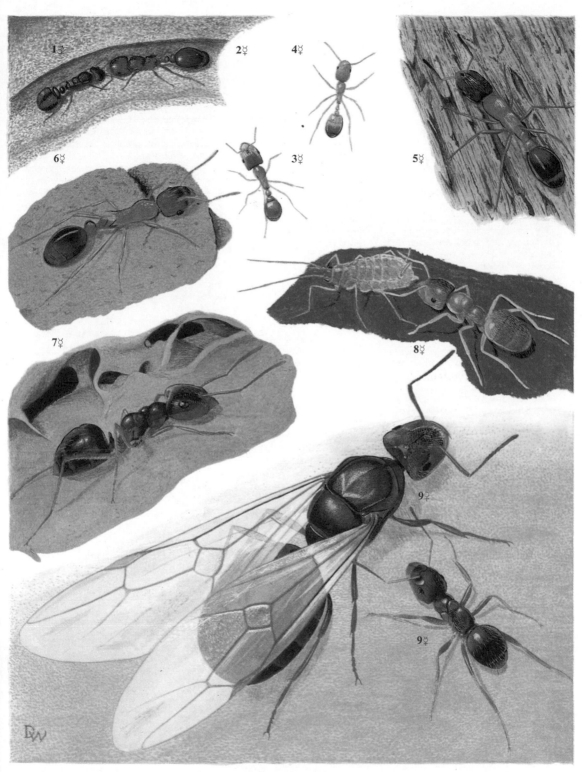

TEN TIMES LIFE SIZE

1 ANERGATES ATRATULUS	2 TETRAMORIUM CAESPITUM	4 STRONGYLOGNATHUS TESTACEUS
6 MYRMICA RUGINODIS	3 MONOMORIUM PHARAONIS	5 LEPTOTHORAX ACERVORUM
7 LASIUS FULIGINOSUS	8 LASIUS FLAVUS	
9 LASIUS NIGER		

1 Wood Ant (*Formica rufa*). This is the ant which makes the familiar ant-hills over much of England and Wales; there are related species which behave in the same way in Scotland and Ireland. The colonies build twig, leaf, and grass-stem domes over their nests (1a) which rise up among bracken on open scrubland or on the bare forest floor. The nests are usually in groups, which are really all part of one community which had grown so large that it needed to spread. They are, therefore, all friendly to each other. Each nest has a number of queens, and a community in course of time produces a population numbering perhaps 100,000 workers. They roam from ground to tree-top in search of insect prey. Separate communities have defined territories for hunting which are jealously defended, and there are often definite paths running from the nest to the trees and shrubs which are in the territory. Wood ants have no sting, but they can squirt formic acid at an enemy, and they also can give quite painful nips with their powerful jaws. The real eggs are white and minute; the large so-called eggs sold as fish or bird food are really the cocoons.

2 Slave-making Ant (*Formica sanguinea*). Also called the Robber Ant, this species differs from other *Formica* in having a notch cut in the middle of the face between the mandibles. Robber Ants live on heathland and other uncultivated places and nest beneath logs or stones or in rotten tree stumps. They are extremely rare, except in the central counties of southern England and in the eastern highlands of Scotland, where they are locally common. They have the interesting habit of making 'slaves' of the workers of other species, especially of *F. fusca* and *F. lemani*. A band of Robbers set out to find a nest of a suitable prey, which they then attack and make off with the worker pupae from the nest. They probably eat some of them, and the rest hatch out and become workers for their captors.

3 Negro Ant (*Formica fusca*). A black, fast-moving, medium-sized ant, abundant over much of the British Isles, though much more common in the south than elsewhere, and sometimes to be found in country gardens. It nests under stones and in old stumps in cultivated places. It differs from *Lasius niger* (p. 151) by its much longer legs and much broader abdominal waist. The worker caste also has three, well-developed, simple eyes.

4 Formica lemani. A black ant, very much like *F. fusca* in habits but distinguished by having a number of upstanding bristles on the upper part of the thorax. It lives further north or at higher altitudes than *F. fusca*, which it more or less replaces in the mountains of the north and west; and it overlaps with *F. fusca* in central England and Wales. Neither of these *Formica* species make twig domes over their nests as Wood Ants do, but they sometimes excavate a great deal of earth, forming mounds on which grass grows.

5 Argentine Ant (*Iridomyrmex humilis*). Although a small ant, its colonies are exceedingly populous, and its ability to over-run new territory, once it has accidentally been introduced, is world famous. Fortunately, it is not able to survive out-of-doors in the British climate, but it does become established from time to time in artificially-heated buildings. Ants, though often a nuisance to man, are not usually seriously harmful; but in the warmer parts of the U.S.A. this ant has become a serious household pest.

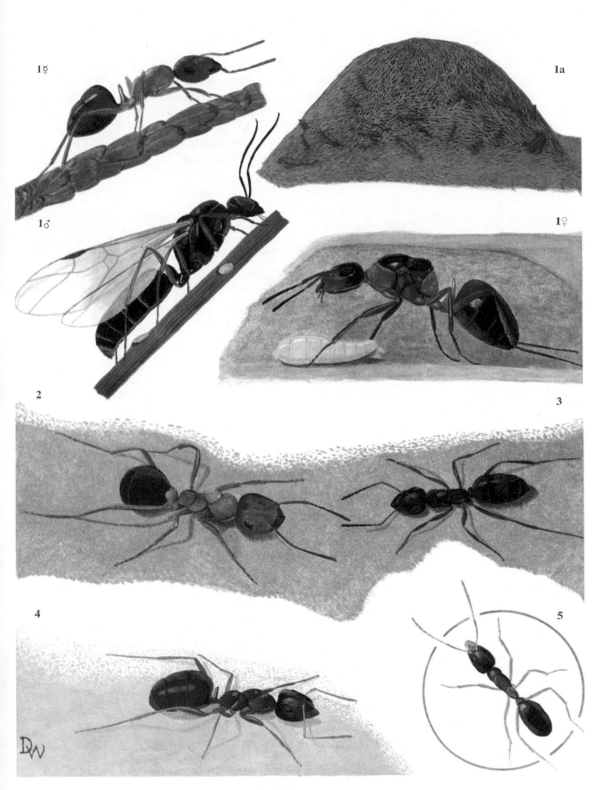

1♀   1a   1♂   1♀   2   3   4   5

SIX TIMES LIFE SIZE (IN CIRCLE × 10)

1 FORMICA RUFA   2 FORMICA SANGUINEA   3 FORMICA FUSCA
4 FORMICA LEMANI   5 IRIDOMYRMEX HUMILIS

The females, but not the males, of all wasps have stings. Like bees, the adults feed on nectar and fruit juices, but they have shorter tongues than bees, and so can suck nectar only from shallow flowers. The larvae feed on the bodies of other insects. Wasps have no pollen-collecting organs. The first two species on this page belong to the Pompiloidea, a group of solitary wasps, the Spider-hunters. The Vespoidea, nearly all black and yellow insects which fold their wings longitudinally when at rest, include the solitary wasps (3–4) and the social wasps (5–7). The third group, the Sphecoidea, which do not fold their wings, are described on p. 156. Nos. 8–9 on this page belong to this group.

1-2 Spider-hunting Wasps (family Pompilidae). There are 40 British species, mostly black with some red on the abdomen. They have large eyes, and long, agile legs, with which they dart across the ground at great speed; but they rarely fly more than a few inches. They catch spiders, which they paralyze with their poisonous stings, and then put in their cells ready for the grubs to devour when they hatch.
Priocnemis perturbator (1) and *Anoplius fuscus* (2) are among the largest spider hunters. They can be distinguished from each other by the amount of red on the abdomen and by the surface of the head, which in *Anoplius* is smooth.

3-4 Potter and Mason Wasps (sub-family Eumeninae). These solitary wasps make no nests as social wasps do, but they either make hard cells out of clay, tiny pebbles, and saliva or burrow in wood or sandy soil. They stock each cell with small paralyzed caterpillars, and then lay an egg before sealing it up.
Eumenis coarctata (3) builds little flask-shaped cells attached to heather stems. *Ancistrocerus parietum*, the Wall-mason Wasp (4), is a common solitary wasp which makes its cells in holes in posts, window frames, crumbling mortar, and such places.

5 Common Wasp (*Vespula vulgaris*). Both this and *V. germanica* are known as the Common Wasp. They have the same habits and are alike to look at, except for some markings. *Vulgaris* has a black 'anchor', and *germanica* three black dots on the front of the head, and the black triangle on the abdomen of *vulgaris* is a diamond in *germanica*. Both are social wasps, making nests in the ground or in attics and outhouses. In late spring, after hibernating, a queen sets to work to make a nest, about the size of a golf ball, out of wood fibre and saliva worked into a pulp. In this is a comb with a single layer of cells, in each of which an egg is laid. When these hatch, the queen works hard feeding them with chewed up flies. In about four weeks, in June, the first workers emerge. Except for egg-laying, they take over all the work of the colony, collecting food, caring for the eggs and larvae, building more layers of cells, and extending the nest until, by August, it may be the size of a football and contain perhaps 2,000 workers.

At the end of the summer a generation of perfect females and males is produced, and these leave the nest and mate. The new queens search for somewhere to hibernate, and the males fly about until they die of cold. The old queen and the remaining workers also soon die. In the spring those queens which have survived the winter start the new colonies.

6 Hornet (*Vespa crabro*). A much maligned wasp which, in fact, is perhaps the most docile as well as being the rarest of British social wasps. It is found only in well-timbered districts, where it nests in hollow trees or sometimes in banks or hanging from the rafters of outhouses. Its reputation comes probably from its size, for the smallest worker is larger than any other wasp queen. Its life history is like that of the Common Wasp. In contrast to Continental Hornets, the top of its head is yellow rather than red. It can be distinguished from queen wasps by its yellow and red rather than yellow and black colouring.

7 Tree Wasp (*Vespula sylvestris*). There are two British species of Tree Wasps, of which this is the more common. It is a little smaller than the Common Wasp, with a much longer face, and at most a black dot on the front of the head instead of an 'anchor'. It nests in hollow trees or occasionally below ground or in roofs. For life history see Common Wasp.

8-9 Sandwasps (*Ammophila sabulosa* and *pubescens*). These look much alike, though they differ in habits. *A. sabulosa* (8), the larger, is often to be seen on sandy heathlands straddled across an enormous caterpillar, which it is about to drag to a prepared nesting hole in the ground, sometimes a considerable distance away. It draws the caterpillar into the hole, lays an egg on it, and then closes the hole, ramming the sand down with a pebble, and smoothing it over so that nothing is seen on the surface. Then the Sandwasp repeats the performance for another egg. *A. pubescens* (9) also provisions her nest with a caterpillar, but instead of sealing up the holes once and for all, she opens them up at intervals to put in another caterpillar. She is the only member of British Sphecoid wasps to continue care of the young.

LIFE SIZE

1 PRIOCNEMIS PERTURBATOR	3 EUMENIS COARCTATA	2 ANOPLIUS FUSCUS
5 VESPULA VULGARIS	4 ANCISTROCERUS PARIETUM	6 VESPA CRABRO
7 VESPULA SYLVESTRIS	8 AMMOPHILA SABULOSA	9 AMMOPHILA PUBESCENS

155

All the wasps on this page and Nos. 8 and 9 on p. 154 belong to the family Sphecidae of which there are over 100 British species. They vary considerably in structure and habits, but they all have a short, collar-like 'neck' between head and thorax, and are coloured black or black with red, yellow, or white markings. They all make their cells in holes in the ground or in woodwork or masonry, and stock them with live insect food for the larvae. They paralyze their victim, usually by stinging it, so that it remains alive, but immobile.

1 **Mournful Wasp** (*Pemphredon lugubris*). An entirely black wasp which nests in beetle holes in old posts and is very common in country gardens. It stocks its cells with aphids.

2 **Mimesa equestris**. This wasp nests in sandy banks, sometimes in quite large communities, though it is essentially a solitary wasp, and the wasps in the community do not co-operate with each other in any way. They stock their cells with aphids and other small Homoptera (pp. 33 – 35). The abdomen is set on a long, narrow 'waist', and the basal segments are red.

3 **Common Spiny Digger Wasp** (*Oxybelus uniglumis*). This is the most common of its genus, which all have wing-like, lateral outgrowths on the thorax and a spine-like growth at the back of the thorax. The abdomen bears yellowish-white patches on the first four segments and usually a complete band on the fifth. It is a wasp of sandy places, and it stocks its cells with small flies (3a). It seems that instead of stinging its prey to paralyze it, it crushes its nerve centres.

4 **Slender-bodied Digger Wasp** (*Crabro cribrarius*). One of the largest of a group of black or black and yellow wasps with large round heads, silvery hairs on the front of the head, and a much reduced wing venation. *C. cribrarius* has yellow spots and bands on the abdomen, and the female thorax has a yellow mark in front and behind. The male has enormously expanded front legs. The cells are made in sandy ground and are stocked with fairly large flies (4a).

5 **White-mouthed Digger Wasp** (*Coelocrabro leucostomoides*). An entirely shining black little wasp, except for pale spurs on the hind legs. It nests in beetle holes in rotten wood, stocking its cells with small flies.

6 **Nysson spinosus**. This is a cuckoo wasp which lays its eggs in the cells of another wasp, *Gorytes mystaceus* (No. 7). It looks very like its black and yellow host, but it has two spines above its 'waist' and much redder legs.

7 **Gorytes mystaceus**. This rather sturdy wasp may be seen extracting frog-hoppers (p. 33) from inside 'cuckoo-spits', for with these it stocks its underground cells. Its head, thorax, and abdomen have narrow yellow stripes, and the male has long antennae, and a considerable amount of yellow on the legs.

8 **Field Digger Wasp** (*Mellinus arvensis*). A late-summer wasp which is abundant in most sandy places from July to late September or even October. It nests beneath ground, stocking its cells with flies. The female is considerably larger than the male. There are various yellowish-white spots on the head and thorax, and the abdomen, especially of the female, is largely yellow. It differs from the various *Crabro* species by having a more complete wing venation and no silvery hairs on the front of its head.

9 **Cerceris arenaria**. One of the two largest British members of the genus which differs from all other wasps in having deep constrictions between the abdominal segments. Both this and the other large species, *C. rybyensis*, are extensively marked with yellow. They are to be seen in July and August on sandy commons in southern England, making vertical burrows, often in very hard ground. All *Cerceris* species prey on small bees or beetles, *C. arenaria* choosing fairly large weevils (9a).

TWICE LIFE SIZE

1 PEMPHREDON LUGUBRIS + APHIDS   2 MIMESA EQUESTRIS + LEAF-HOPPER

3 OXYBELUS UNIGLUMIS + FLY   4 CRABRO CRIBRARIUS + FLY

6 NYSSON SPINOSUS   7 GORYTES MYSTACEUS

5 COELOCRABRO LEUCOSTOMOIDES   8 MELLINUS ARVENSIS + FLY   9 CERCERIS ARENARIA + WEEVIL

In Britain there are about 250 different species of bees (Apoidea). A few are social insects (*see* p. 162), some have a cuckoo-like habit of laying their eggs in other bees' nests, and the majority are solitary, sometimes with rudimentary social behaviour. Nearly all bees are very hairy, many with featherlike hairs, which distinguishes them from hairy wasps. All bees feed on the nectar of flowers, and feed the larvae on pollen mixed with honey, and the workers of all except the 'cuckoo-bees' can collect and carry pollen back to the nest in hairy hollows in the hind legs (pollen baskets) or long hairs under the abdomen. Most bees have long 'tongues', actually extensions of the lower lip, for sucking nectar from flowers.

The Mining Bees (Nos. 2–8) look rather like Honey Bees, though of the 100 or so British species, most are much smaller. They have shorter 'tongues' than most bees, and most prepare cells for their eggs by mining in sand or earth. The female digs the burrow, sometimes 2 feet deep. She pulls the earth backwards with her fore feet and brushes it out with her hind feet, scattering it round so that the entrance is not easy to see. She makes a cell at the end of the tunnel and provides this with a ball of honey and pollen. She has to make six or seven journeys to flowers, each time carrying half her own weight of pollen. Then she lays her egg on the ball and closes the cell. Next she makes another cell higher up the tunnel and provisions that in the same way. She makes three to six cells; then she comes to the surface and dies.

1 Colletes daviesana. The members of this family have short, forked, wasp-like mouthparts. The 8 British species are all much alike, with very hairy head and thorax and bands of white hairs on the abdomen. They nest in burrows 8 or 10 inches deep in hard, sandy soil, and also side by side in soft mortar and decaying masonry. They lay their eggs in papery cells, which they fill with almost liquid honey, ready for the larvae.

2-3 Halictus. There are 35 British species of this genus, the females of which all have a hairless ridge running down the end part of the abdomen. *H. xanthopus* (2) is one of the largest and brightest coloured — most are rather dingy. It has golden-coloured legs and its abdomen is oval-shaped. *H. calceatus* (3) is perhaps the commonest of the more insignificant ones. Several segments of the male's long and narrow abdomen are reddish, and the female's abdomen is more egg-shaped, her legs are black, and the pollen baskets on the hind legs have golden hairs.

4 Sphecodes gibbus. These black and red 'cuckoo bees' lay their eggs in the cells of Halictus and some other mining bees. Their almost hairless bodies have no arrangement for collecting pollen. The 16 British species are almost impossible to tell apart, and all vary a good deal in size.

5-7 Andrena. All the 60 or so species of this genus of mining bees are excellent pollinators of fruit trees, and their stings are too weak to penetrate the human skin. So they are entirely beneficial. apart from their habit of nesting in lawns and making them look untidy. *A. haemorrhoa* (5) is one of the commonest and earliest to appear. The female has redder hair on the thorax than the male, and the golden hairs at the tip of the abdomen

are redder. The male is brown haired with paler legs. *A. flavipes* (6) appears rather later in the spring. The females sometimes form large colonies of burrows close together, and in midsummer a second generation appears. The female's abdomen is almost bare, except for bands of whitish hairs; but the male's abdomen is covered half with brownish and half with black hairs, with bands of paler hairs at each segment.

A. armata (7) is often called the Lawn Bee because many nest together in lawns, throwing up little conical piles of soil. There is only one generation, and the bees have disappeared by the end of June. *A. armata* is abundant, and the thick, bright-chestnut hairs on the female's body are conspicuous. The more slender male has a ragged, thinner covering of pale-brown hairs.

8 Dasypoda hirtipes. A summer mining bee of coastlands. The female has very large, hairy pollen baskets on the hind legs, but otherwise neither male nor female is unusually hairy. They are buff-coloured with black abdomens, except for a buff base and bands of whitish hairs.

9- Flower Bees (*Anthophora retusa* and *acervorum*). Both
10 these species are very hairy, stout bees, among the earliest to be seen in the spring, and can easily be mistaken for Bumble Bees. The females are blacker than the males and have orange hairs on the pollen baskets of the hind legs. The males have long, hairy legs, *Anthophora acervorum* (10) having brushes of very long, black and grey hairs on the middle pair. Both species nest in the ground, and *A. acervorum* will also burrow in walls. The cells are little clay 'pots' which the females fill with honey and pollen. They then lay an egg on the honey and close the pot — hence their other name, Potter Bees.

TWICE LIFE SIZE

1 COLLETES DAVIESANA	2 HALICTUS XANTHOPUS
4 SPHECODES GIBBUS	3 HALICTUS CALCEATUS
5 ANDRENA HAEMORRHOA	6 ANDRENA FLAVIPES	8 DASYPODA HIRTIPES
9 ANTHOPHORA RETUSA	7 ANDRENA ARMATA	10 ANTHOPHORA ACERVORUM

LEAF-CUTTER BEES,
CUCKOO BEES, ETC.

The family Megachilidae (Nos. 5–9) consists of bees which collect pollen on the belly, not on the hind legs. The species of the genus *Megachile* (5) are leaf-cutters. Cuckoo bees (Nos. 2–4) are those that use the work of other bees to rear their young, which they cannot do themselves for they are without any pollen-collecting equipment. They belong to several different families (*see* also p. 162), and are often closely related to their hosts, though they may not look very like them.

1 Eucera longicornis. This large bee is the host of the Cuckoo bee *Nomada sexfasciata*. The male has very long antennae which reach back to the tip of the abdomen. The female has shorter antennae, no yellow on the head, and, apart from a golden-haired tip, a darker abdomen. The segments of the abdomen are marked by bands of pale hairs. This bee is local and restricted to the southern part of England. It nests on the ground, often in large colonies, between May and July.

2-4 Cuckoo Bees (genera *Melecta* and *Nomada*). These are sometimes called 'Homeless Bees' because they make no nests but, like cuckoos, lay their eggs in other bees' nests.
Melecta punctata (2) is one of a pair of large, grey Cuckoo Bees which prey on *Anthophora* bees (*see* p. 159), but are far less common than their hosts. They are armed with a pair of spines on the thorax.
The genus *Nomada* consists of 27 British species of wasp-like, hairless bees, unlike the species of *Andrena* (*see* p. 159) on which they prey. Like their hosts, they have feeble stings and rely on stealth, rather than force, to enter their hosts' burrows. There, the female lays her egg in the host's cell, and the grub, when it hatches, feeds on the food provided and, later, on the host's egg or larva as well. *N. marshamella* (3) is black and yellow, sometimes with a little reddish-brown on the abdomen. *N. ruficornis* (4) is brighter coloured, with red, black, and yellow stripes, and the female has red stripes on the thorax.

5 Megachile centuncularis. This typical 'leaf-cutter' bee makes its nest in holes in wood, and collects pollen in bright orange hairs on its belly. The female cuts semi-circular notches in leaves, especially rose leaves, carries these to the nest, and rolls them up into a hollow cylinder. She seals the bottom with a piece of leaf, puts

in a supply of honey and pollen, lays an egg, and then seals off the cell with another bit of leaf. Then she supplies a second cell on top, and so on until the cylinder is full. She then leaves the burrow and dies. When new bees emerge, they eat their way out, the topmost one first. The first to emerge are always males, and these wait on nearby flowers until the females appear.

6 Anthidium manicatum. A bee often called the 'hoop shaver' or 'wool carder' because of its habit of stripping the woolly coating off the stems and leaves of various plants. It rolls up the 'wool' into a ball, carries it to the hole, and lines the nest with it to protect the almost transparent, membranous cells. This rather hairless bee has various yellow spots and stripes on its body, and, unlike most bees, the male is larger than the female.

7-8 Osmia rufa and **bicolor.** These are the largest of the British *Osmia*, some being only half the size. Some have metallic blue or green abdomens. *O. rufa* (7) is a familiar insect, often nesting in nooks and crannies in buildings. The female differs from the male in having black instead of white hairs on the face and two large, forwardly-directed horns on the face. *O. bicolor* belongs to chalk downlands, where it nests in old snail-shells. The male differs from the female in being more brownish-yellow.

9 Chelostoma florisomne. A little black bee which, apart from the feathery, pollen-gathering hairs on its belly, might be mistaken for a little wasp. The males have unusual belly segments with which they hold the female during mating. The females nest in beetle-holes in old posts, and the males spend the nights and also dull days sleeping, curled up in flower heads.

TWICE LIFE SIZE

1 Eucera longicornis   2 Melecta punctata
3 Nomada marshamella   4 Nomada ruficornis   5 Megachile centuncularis
6 Anthidium manicatum   9 Chelostoma florisomne
7 Osmia rufa   8 Osmia bicolor

These bees are the social bees, which live in organized communities, though the Cuckoo Bees, like those described on p. 160, are parasites on their relatives and do no work themselves.

1-4 Bumble Bees (genus *Bombus*). These differ from the Honey or Hive Bees in that the colony is annual, and no store of honey is laid up for the winter. Only the young fertilized queens survive the winter, and each starts up her own colony the next year. They are larger and much more colourful than Honey Bees, and the castes are much alike, except that the queens are the largest.

The fertilized queen, having hibernated in some protected place such as in a hole or under moss, comes out in the spring, gathers some nectar, and then starts to found a family. Having located a suitable nesting place, above or below ground according to her species, she makes her nest of, for instance, grass, moss, or leaves, and then makes 'pots' of wax and pollen, into which she lays the first batch of eggs. When these hatch, she continues to provide them with honey, while at the same time she is laying up storage cells for honey and more cells for further eggs. After about 3 weeks, the first young bees emerge. These are always workers, that is infertile females. They now take on the work of honey-gathering and cell-building, while the old queen concentrates on egg-laying. *Bombus* cells are rounded and arranged in irregular clumps, and the nest is not nearly so orderly as that of the Honey Bee.

In course of time both females and males are produced, as well as many more workers, and by then a large colony may contain several hundred bees. Towards the end of the summer the males and females fly out and mate. The males, which are not allowed to re-enter the nest, soon die. The females start searching for a safe place to hibernate, and the old queen and workers die when the cold weather comes on.

There are 16 British species of the genus *Bombus*. The earliest to appear in the spring is *B. pratorum*, a rather small Bumble Bee with a red tail and a good deal of yellow on the body, especially the male. Its nests are usually below ground. *B. lapidarius* (1), a black bee with a red tail, usually nests deep down a mouse hole, and makes larger colonies than any other British species, often containing several hundred workers. *B. terrestris* (2), the largest, also nests below ground. The queen has a gingery tail, but those of the males and workers are white.

B. agrorum (3) and *B. muscorum* (4) both nest above ground and are sometimes called 'carder-bees'. They make their nests largely of grass and moss, not unlike those of small birds. The large *B. muscorum* is usually found in marshy places, and a peculiar reddish form, with black hairs underneath, is found on some west-coast islands.

5-7 Cuckoo Bees (genus *Psithyrus*). These attack the nests of young Bumble Bee queens, kill them, and use their workers to rear their own young. The hibernating females emerge after the *Bombus* females to allow for these to have produced enough workers, for the *Psithyrus* queens are unable to produce more workers nor can they collect pollen for themselves. Some Cuckoo Bees, such as *P. rupestris* (5), look very like their hosts, in this case *B. lapidarius*, though the females do not have the pollen-collecting arrangements. *P. vestalis* (6) attacks and resembles *B. terrestris*, but instead of a yellow band, it has two yellow patches on the abdomen, which are less distinct on the male. *P. campestris* (7) does not at all resemble any of the several brown Bumble Bees which it attacks. It normally has bands of yellow on the thorax and also some yellow on the abdomen; but very dark or even black individuals are not uncommon.

8 Honey Bee (*Apis mellifera*). This is the Hive Bee, the only British species of its genus; it is almost entirely domesticated. Honey Bees form permanent communities in a hive or, in the wild state, in a hollow tree. Enough food is stored through the summer to keep the community alive through the winter. The community consists of drones (males), whose sole purpose is to fertilize the queen and who die after doing so, or are killed by the workers; a queen (female) who lays eggs; and workers (sterile females). Eggs to produce queens are laid in large, rounded cells (one is shown on the plate), and the larvae are fed on richer food (royal jelly) than the worker larvae, which develop in typical hexagonal comb-cells. Only the workers have pollen- and honey-collecting arrangements on the hind thighs and wax-secreting organs. They do all the work of the hive, feeding the queen, drones, and larvae, cleaning the hive, keeping off intruders, building up the cells, and collecting honey and pollen. The queen continues to lay an egg in each cell until, after 3 or 4 years, she is too old, when she is ruthlessly killed by the workers. Usually about May the hive is becoming overcrowded, and then, when a new queen is about to emerge, the original queen will fly out of the hive, taking with her almost half the community, perhaps a swarm of some 30,000 bees. When she alights, the swarm settle on top of her, and at this point the bee-keeper 'takes the swarm' and introduces them to a new hive, where they proceed to build up fresh combs of cells.

In the meantime the new queen in the old hive, immediately on emerging, stings to death all other queens in their cells and, after a mating flight, proceeds with the business of egg-laying.

TWICE LIFE SIZE

1 Bombus lapidarius   3 Bombus agrorum   2 Bombus terrestris
5 Psithyrus rupestris   4 Bombus muscorum   6 Psithyrus vestalis
8 Apis mellifera   7 Psithyrus campestris

163

GROUND BEETLES (*Carabidae*) 1.

This very large family of mainly carnivorous beetles (see also p. 166) have long, powerful legs adapted for running down their prey, and thread-like antennae. They have five joints to all their feet, in contrast to the False Ground Beetles (*see* p. 180). The sexes are usually similar.

1-2 Tiger Beetles (sub-family Cicindelinae). These have relatively enormous, powerful mandibles, which over-lap when not in use. They fly and run actively in the sunshine and have a 2-year life cycle. A common bettle on dry heaths and sandy places is *Cicindela campestris* (1), the Common or Green Tiger Beetle. It is active on sunny days in spring and early summer. In late summer, the larva, which has huge mandibles, digs a burrow up to 12 inches deep in a bare patch of ground, at the top of which it anchors itself by two hook-like spines. Thus, with head and mandibles filling the small entrance hole, it awaits its prey (1a). As soon as an ant or other insect comes near enough, the larva shoots out its mandibles, seizes the prey, and drags it to the bottom of the burrow to be devoured. When fully grown after two summers, it digs a pupal chamber at the bottom of the burrow, blocks the entrance with soil, and then pupates; the adult beetle remains in the chamber until the spring, when it bores its way to the surface.

The Wood Tiger Beetle (*Cicindela sylvatica*) (2) is usually larger than the Common Tiger Beetle, and has large cream-coloured markings on its dark wing-cases; it has a similar life history. It seems to be restricted to sandy heaths in southern and eastern England, and is only locally plentiful.

3 Calosoma inquisitor. A predatory beetle of oak woods where, both as larva and adult, it pursues the larvae of other insects among the trees, particularly the destructive Green Oak-roller (p. 121) and the Winter Moth (p. 103). Though widespread in England and Wales, it is local and rather uncommon. It has also been found in eastern Ireland. It flies readily in May and June.

4-6 Ground Beetles (sub-family Carabinae). These have less formidable mandibles than the Tiger Beetles. They are mostly dark-coloured or metallic, and nocturnal, hiding under cover by day, They hunt on foot, and many species cannot fly. When disturbed, the larger ones may discharge in defence an acrid liquid from their anus, and sometimes from their mouths as well. Many hibernate either as larvae or adults, but some are about in the winter; they are most active between April and September. The larvae look much alike, except in size, and usually pupate in earthen cells in the soil.

The Violet Ground Beetle (*Carabus violaceus*) (4) is a large, flightless beetle with violet or purple borders to its otherwise black thorax and wing-cases. It is found almost everywhere, especially from June to August, but is local in Ireland. By night it attacks and devours animals such as earth-worms and many insect pests; by day it lurks under moss, bark, stones, and logs. The larvae behave in much the same way as the adults, and they hibernate as larvae.

Another beetle of the same genus is *C. granulatus* (5) which varies from dark copper-brown to greenish-black in colour, with three ridges separating the rows of large oblong granules on each wing-case. The wing-cases are not fused, and long vestigial wings lie beneath. This beetle inhabits marshy places such as the margins of ponds with decaying willows, and its habits and life history are similar to the Violet Ground Beetle. *C. nemoralis* (6) is fairly common, and is particularly active at night in August and September. It is black with three rows of large, shallow, metallic depressions on its dark greenish or bronze-coloured wing-cases, and purple-pink edges to its thorax.

7 Nebria brevicollis. This and the remaining beetles on this page have no common names. This is a very common, flattish, shiny black beetle found almost everywhere, both as adult and larva, in damp, shady places, where it lurks under stones or bark in the daytime. It is to be found throughout the year, but most often in spring and autumn. The larva is very active and carnivorous. A similar but smaller bright-blue beetle, *Leistus spinibarbis*, is quite common in similar places.

8 Notiophilus biguttatus. A common little beetle which runs swiftly in the sunshine in woods, gardens, and open country. It has the characteristic broad, arrowed head, large eyes, and shiny bronze colour of its genus.

9 Elaphrus cupreus. Both in behaviour and appearance this resembles a very small Tiger Beetle. It runs very fast in the sun in wet places, such as the muddy edges of lakes and rivers, and, like the smaller, greener *E. riparius*, is found locally over most of Britain.

10 Loricera pilicornis. A common, brassy beetle, with distinctive long hairs on its antennae. It lives amongst dead leaves and tree-roots, especially along the margins of rivers and streams, almost everywhere except for the extreme north. The active, predatory larvae have powerful mandibles.

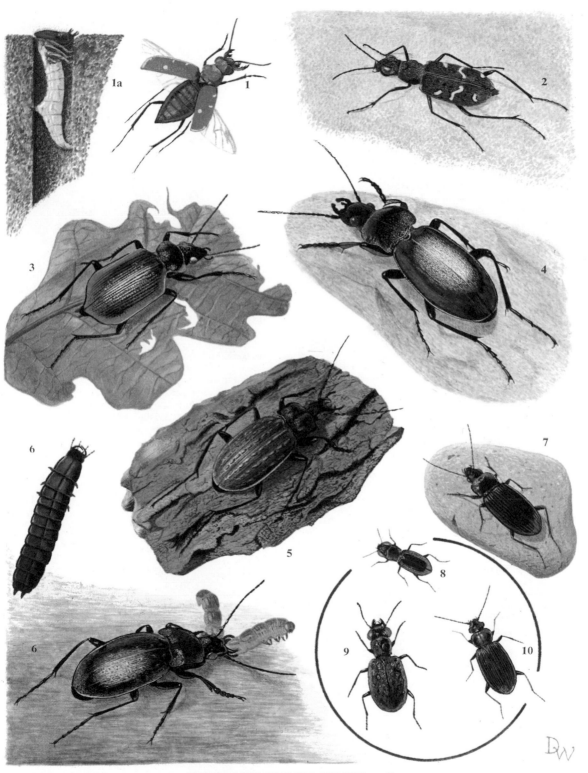

TWICE LIFE SIZE (IN CIRCLE × 3)

1 CICINDELA CAMPESTRIS 2 CICINDELA SYLVATICA
3 CALOSOMA INQUISITOR 4 CARABUS VIOLACEUS
6 CARABUS NEMORALIS + LARVA 5 CARABUS GRANULATUS 7 NEBRIA BREVICOLLIS
8 NOTIOPHILUS BIGUTTATUS 9 ELAPHRUS CUPREUS 10 LORICERA PILICORNIS

There is a general note on Ground Beetles on p. 164. The first species belongs to the sub-family Carabinae, and the rest to the large sub-family Harpalinae, which in most respects resembles the Carabinae (p. 164), but has a notch on the inner side of the tibia of the foreleg. They are nocturnal and hide by day under cover such as stones, and some are small and difficult to identify. None of the species on this page have common names.

1 **Clivina fossor.** A small beetle, fairly common throughout the British Isles, apart from the Scottish Highlands, and distinct from the less common *C. collaris* by its black instead of reddish-brown wing-cases. Both species have the thorax divided from the abdomen by a waist, and their forelegs are modified for digging (1a). They hide by day in damp places under dung or stones or in garden refuse.

2 **Trechus quadristriatus.** A very small, reddish-brown ground beetle, very common in sandy localities throughout the British Isles, except Ireland, where it is rather scarce. It possesses fully-developed wings.

3 **Bembidion guttula.** A very small beetle, common throughout the British Isles, which is black with a small, pale spot near the hind end and reddish legs. It is much like the also very common *B. lampros*, which is a shiny bronze colour and has rather redder legs. There are many other species, some with light markings on the wing-cases.

4 **Harpalus affinis.** A very common beetle found almost everywhere. It varies a great deal in colour; but metallic shades of bronze or bright green are usual. The antennae and legs are normally red, and the female is generally duller than the male. It is chiefly nocturnal, hiding by day in burrows under stones or roots, where the larva also lives; but beetles may often be seen about in the sunshine during spring and early summer.

5 **Badister bipustulatus.** A common beetle of both wet and dry places over most of the British Isles, though rare in the Scottish Highlands and Ireland. It frequents, among other places, the roots of willows and other trees and hides by day under stones, rotten wood, and moss. It is like the rarer *B. unipustulatus* but is smaller, with a narrower black head, and the sides of the red thorax broader behind.

6 **Anisodactylus binotatus.** A generally black beetle which hides by day under lumps of earth and stones in damp places in most parts of the British Isles; it is commonest in southern England and scarcest in Scotland and Ireland. It is similar to the rarer *A. poeciloides*, which is found only in salt marshes and is usually metallic green or coppery and rarely black.

7 **Amara aenea.** A somewhat oval-shaped ground beetle, which may be brassy, and is occasionally greenish or bluish, with the basal segments of the antennae red. It is very common over all the British Isles, except for Ireland and the extreme north of Scotland, and can often be seen, after hibernation, running in the spring sunshine on paths or bare ground. It flies readily and, when interfered with, discharges an evil-smelling, pungent fluid.
Beetles of this genus are partly vegetarians, eating seeds and other parts of plants, and partly carnivorous, attacking the larvae of small insects. The larvae hatch from eggs laid in the soil or under stones, hide underground during the day, and come out to feed at night. After 6 to 8 weeks they pupate, the beetles emerging in about a month.

8 **Pterostichus cupreus.** The general colour of this handsome beetle ranges from brassy or coppery to green, purplish, or almost black. The two basal segments of the antennae are red, and the legs are usually black, though a red-legged variety also occurs. Both larvae and adults hide under stones and lumps of earth by day, emerging at night to feed. *P. cupreus* is locally common throughout the British Isles, except for the Scottish Highlands.

9 **Pterostichus nigrita.** A rather slim beetle, common in marshes all over the British Isles. It spends the day under stones, lumps of earth, or logs, or among the roots of plants, and emerges at night to hunt its prey, as does also its larva.

FOUR TIMES LIFE SIZE

1 CLIVINA FOSSOR   2 TRECHUS QUADRISTRIATUS

3 BEMBIDION GUTTULA   4 HARPALUS AFFINIS   5 BADISTER BIPUSTULATUS

6 ANISODACTYLUS BINOTATUS   7 AMARA AENEA

8 PTEROSTICHUS CUPREUS   9 PTEROSTICHUS NIGRITA

The first three species on this page are ground beetles of the sub-family Carabinae (*see* p. 164). The rest are water beetles, as are the first two species on p. 170. Typical water beetles, such as the Dytiscidae (Nos. 5 – 8) spend almost all their lives in water, and, if the water dries up, they either burrow into damp mud or fly elsewhere. Most have a supply of air enclosed in their wing-covers to use when under water. This makes them so light that they cannot stay below water except by swimming with their hind legs or gripping weeds or stones with the middle pair. The front pair are used for catching aquatic animals on which most of them feed. They renew the air supply fairly frequently by projecting their tail-ends above the water.

1 Agonum marginatum. A ground beetle which frequents muddy places by water, hiding by day among roots and moss, and under stones. It is rather common locally in southern England, Ireland, and Wales, but scarcer further north up to the south of Scotland.

2 Demetrias atricapillus. A very common little ground beetle often found on nettles and low herbage, throughout England and Wales, but rare in the extreme north, unknown in Scotland, and widespread in Ireland. It is often paler than the one illustrated.

3 Bombardier Beetle (*Brachinus crepitans*). This small beetle defends itself by ejecting with explosive force from its anus a jet of caustic liquid, which at once volatilizes and looks like a little puff of smoke. Bombardier Beetles are most usually to be found in August under roots and stones in southern England, especially in chalky districts, along the south coast and on river banks, where they may be abundant, several under one stone. They are found locally in south Wales and south-east Ireland. The wing-cases may be blue-black, greenish-black, or black. The larvae are like those of other ground beetles.

4 Screech Beetle (*Hygrobia hermanni*). This sole British member of the family Hygrobiidae, sometimes called the Squeak Beetle, squeaks loudly when disturbed by rubbing the tip of the abdomen against a file under each wing-case. It lives in ditches and ponds in England and south Wales, being fairly common in the south-east but quite rare in the north. It feeds on insect larvae and worms from March until late autumn, when it hibernates, presumably in the mud. The larvae live at the bottom of the pond, breathing through gills, and appear to feed only on *Tubifex* worms. They pupate out of the water, the whole life-cycle taking from 9 to 15 weeks.

5 Platambus maculatus. A locally common beetle in running streams over most of the British Isles, except Ireland. The eggs are probably laid among aquatic plants in mid-summer; the larvae overwinter; and the adult beetles appear in May or June, but may be found in most months. The colour varies to almost all black.

6 Agabus bipustulatus. A very common beetle in stagnant water all over the British Isles. The female is duller than the male, with more densely and finely ridged wing-cases. The larvae generally hatch in the autumn, overwinter, and pupate in the spring in cells constructed in the mud at the water's edge. The adults emerge from May to September and are found throughout the year.

7 Great Diving Beetle (*Dytiscus marginalis*). This large beetle, fairly common in ponds and stagnant waters throughout Britain, may be 1½ inches long, and is to be found all the year round. The duller females deposit their eggs in water plants during the summer, and the larvae pupate the following spring in mud cells at the water's edge. When full grown, the larvae are 2 inches long, with huge mandibles, and are even more rapacious than the adults: both attack almost any living creature within their power, especially tadpoles and small fish. The beetles, which may live up to 3 years, fly readily after dusk and soon colonize new waters. The females of both this and the next species have grooved wing-cases, and the males have large, round suckers on the fore feet.

8 Acilius sulcatus. A widespread beetle active from March until November in ponds on sandy, gravelly, or peaty ground. The shrimp-like larvae feed in the water throughout the summer on small aquatic animals and, when full-grown, construct pupal cells in the mud at the water's edge. They usually overwinter as adults.

9- Whirligig Beetles (family Gyrinidae). These small,
10 shiny black beetles have a habit of gyrating in groups on the surface of stagnant or slow-moving water. When alarmed, they dive rapidly. They have two pairs of eyes for seeing both above and below the water, legs well adapted for swimming, and abnormal antennae.
Gyrinus marinus (9) is found in open water at the edges of lakes, canals, and rivers over most of Britain, often in large groups. They are active, usually by day, from spring to autumn, and appear to feed on the surface on dead insects and even dead fish. The eggs are laid on submerged plants, and the larvae stay on the bottom of the pond until nearly full grown. They pupate in cocoons of mud or other material attached to waterside plants.
The Hairy Whirligig (*Orectochilus villosus*) (10) lives in running water and hides by day. It gyrates on the water surface at night from June to September. The eggs are laid on the roots of water plants, and the larvae live among the gravel and stones in shallow water.

LIFE SIZE (IN CIRCLES × 3)

1 AGONUM MARGINATUM   2 DEMETRIAS ATRICAPILLUS   3 BRACHINUS CREPITANS
4 HYGROBIA HERMANNI   5 PLATAMBUS MACULATUS
7 DYTISCUS MARGINALIS + LARVA   9 GYRINUS MARINUS
6 AGABUS BIPUSTULATUS + LARVA   8 ACILIUS SULCATUS   10 ORECTOCHILUS VILLOSUS

The first two species on this page are water beetles, and the rest are scavengers and predators belonging to the family Silphidae. The burying beetles (Nos. 3 and 4) are unusual in the insect world in that they provide for their young, as do some of the dung beetles (*see* p. 180). The carrion beetles (Nos. 5, 6, and 7), except for the Large Carrion Beetle, are smaller than the burying beetles, shorter, flatter, and more oval and compact, with shorter, weaker legs, and much less knobbed antennae. In the illustration they are shown twice natural size. Most Silphidae are good fliers, as their way of life requires.

1-2 Vegetarian Water Beetles (*Hydrophilidae*). In contrast to the water beetles on p. 169, the larvae only are carnivorous. The beetles have short, narrowly clubbed antennae with fewer than the normal eleven joints, and palpi (mouth feelers) often longer than the antennae. They are mostly poor swimmers and renew their air-supply differently from the diving beetles — by protruding their antennae, usually one at a time, through the water surface until the air makes contact with the bubble of air round the beetle's underside. The larvae take in air through a pair of spiracles in the tail. The females of many species carry their eggs in silken cases attached to the abdomen or hind legs.

The Common Hydrobe (*Hydrobius fuscipes*) (1), found over most of Britain in stagnant water, fixes the egg-case to a leaf or stem of a water plant just below the surface, with a ribbon of silk, as shown in the picture. The larva holds its prey out of the water to eat it. The Great Silver Water Beetle (*Hydrophilus piceus*) (2), the bulkiest British Beetle, and nearly 2 inches long, swims better than most of its family; it lives in marsh dykes in a few parts of southern England, and used to be far more common than it is now. The large silk-cases holding 50–60 eggs are fixed to plants, with a hollow 'mast' sticking out of the water. The big, voracious larvae, on hatching, swallow air enclosed in the case and so become buoyant. Out of water, an alarmed larva may throw back its head and hiss. It holds its prey, usually water snails, in the hollow of its back, and bends its head over backwards to reach it with its powerful jaws, specially shaped for this purpose. It pupates in a large cell in the earth at the water's edge. The beetle flies at night and is attracted by bright lights.

3-4 Burying Beetles. The seven British species vary a good deal in size. The two shown here, the Black Burying Beetle (*Necrophorus humator*) (3) and the Sexton Beetle (*Necrophorus vespillo*) (4) are both common in the south, but the latter is rarer in the north and in Scotland. The Black Burying Beetle is black except for the orange clubs on the antennae — the only British species, apart from one very rare one, to be without the reddish-orange bands on the wing-cases. The Sexton Beetle, the smaller of the two, is distinguished by the curved tibiae (shanks) of the hind legs and by a fringe of yellow hair on the front margin of the thorax.

Several burying beetles together, having found the dead body of a small animal, proceed to bury it. They scrape away the soil beneath it, or, if the ground is too hard, they shift it to more suitable ground by getting underneath it, turning on their backs, grasping it with their jaws, and pushing. They cut away any impeding grasses or roots and even sometimes gnaw off a limb of the corpse to free it. As soon as the corpse is buried and nothing shows but a slight mound, each female makes a ball of well-masticated carrion in an underground chamber and lays her eggs in a short tunnel leading out of it. She stays by this ball, which the grubs gradually consume, helped by the female's regurgitated food until they are independent. Later, they migrate to the buried carcase. In this way they grow more quickly and are fully grown before the food is exhausted. In about two weeks they burrow deeper and pupate. The adults, when not working and in winter, rest in shallow burrows. Those that succeed in surviving the emergence of their young apparently kill each other off.

5-7 Carrion Beetles. Most but not all these feed on carrion, but they do not bury the carrion. They find it by its smell and lay their eggs on it. Their larvae are rather like woodlice, black and shiny above, and are more active than burying beetle larvae. Some carrion beetles, such as the broad *Oeceoptoma thoracicum* with its brick-red thorax, feed on putrid fungi as well as carrion; others attack caterpillars, snails, and also plants, perhaps eating mainly or even solely the maggots present in putrefying matter.

The Rough Carrion Beetle (*Thanatophilus rugosus*) (5) is a true carrion feeder. It used to be found throughout Britain, but is now replaced in many districts by *T. sinuatus*, which is similar but lacks the wrinkles on the wing-cases.

The Four-spot Carrion Beetle (*Xylodrepa quadripunctata*) (6) lives mainly in woods on oak trees where, in May and June, it hunts caterpillars such as the Green Oak-roller larvae (p. 120). It is not very common, except in certain years and in some places such as Epping Forest. It is rare in the north and in Ireland.

The Smooth Carrion Beetle (*Ablattaria laevigata*) (7) feeds on snails, for which its slender, elongated head and stout legs admirably fit it. It is local over England and Wales, being found in chalk and limestone districts, often near the coast. It is nocturnal, hiding by day under stones, as does the very similar and more common *Phosphuga atrata*, which is shinier and rougher, with longer and thinner legs and antennae.

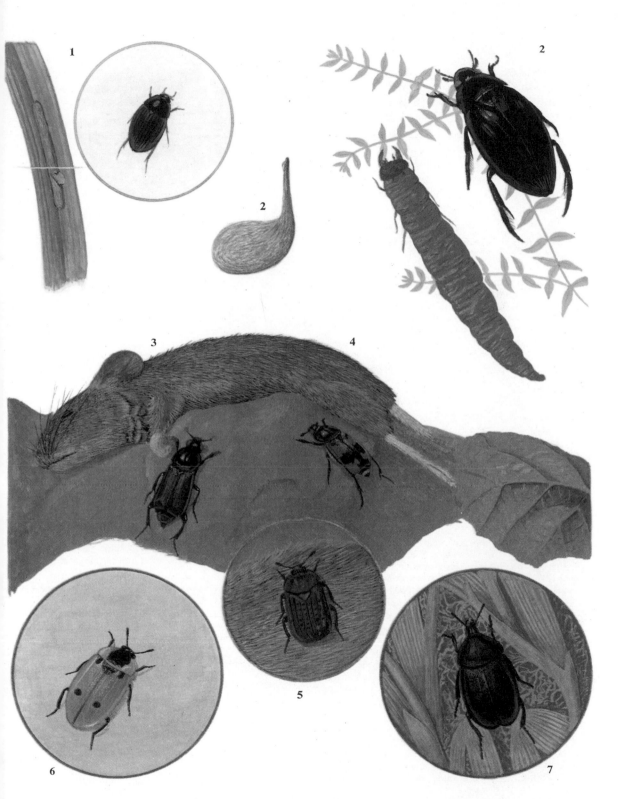

LIFE SIZE (IN CIRCLES × 2)

1 HYDROBIUS FUSCIPES + LARVA   2 HYDROPHILUS PICEUS + LARVA AND EGG CASE

3 NECROPHORUS HUMATOR   4 NECROPHORUS VESPILLO

6 XYLODREPA QUADRIPUNCTATA   5 THANATOPHILUS RUGOSUS   7 ABLATTARIA LAEVIGATA

The Rove Beetles are an immense family, the Staphylinidae, of which there are more than 900 British species, most of them very small and without common names. They have short wing-cases but well-developed wings, and they are in general long and narrow. They vary considerably, not only in size but also in habits and habitats, but most are scavengers and carnivorous, both as larvae and adults, and are often to be found where there is decaying matter, either plant or animal. The larvae are like the larvae of ground beetles, but the pupae are fairly hard, red-brown, rigid, and smooth, with a flattened upperside. The larger species pupate underground.

Most Rove Beetles move quickly and run fast. Many fly by day in warm weather, but some are nocturnal. They are often called 'cocktail beetles' because of a habit, when molested, of cocking their tails over their backs (*see* No. 6) and spurting a little strong-smelling vapour in the face of the enemy. This movement also helps them to fold their long wings under the small wing-cases — a feat which they achieve with wonderful ease.

1-2 Oxytelus laquaetus (1), **O. tetracarinatus** (2). These both belong to a genus of small, flattened, mostly shiny and rather slow-moving beetles, with a ridged and furrowed thorax. The males have broader heads and smaller eyes than the females. *O. laquaetus* is one of the larger species and is common in and under cow-dung. The small *O. tetracarinatus* is very common, not only in dung but in heaps of any rotting herbage and also on the wing. Its legs are often darker than shown in the picture.

3 Stenus cicindeloides. There are over 70 British species of this genus, all small and dark, with large, prominent eyes, giving them an almost hammer-headed aspect. The antennae have slender clubs. Most are found in damp or marshy places. *S. cicindeloides* frequents seed-beds in the south and midlands, and is sometimes to be found in winter, with others of the genus, in the dead stems and debris of reeds.

4 Paederus littoralis. A wingless beetle with striking colours, the wing-cases sometimes being green rather than blue. It is active by day and is not uncommon, especially on chalk, crawling in the open amongst grass, stones, or rubbish, or low down on tree-trunks. It sometimes climbs up hedges. The other three species of this genus have wings and are found in marshy places.

5 Philonthus politus. Most of the 50 or more British members of this genus are medium-sized, active beetles which hunt small creatures in dung, carrion, rotting fungi, or at grass roots and in moss. *P. politus* is common in the first three of these habitats. It is one of the larger species, but is not easy to distinguish from several others.

6 Devil's Coach-horse (*Ocypus olens*). The best-known and one of the largest of the family, to be found in all parts of Britain but most frequently in open country, including cultivated land. It is mainly nocturnal, spending the day under stones, loose bark, among the roots of plants, or any other cover; but it may well be seen on fine days running on paths. Both beetle and larva are very fierce, attacking anything which resists them and giving sharp bites. When alarmed, the beetle opens wide its jaws and raises its tail over its back in a defensive attitude. The closely related and handsome genus *Staphylinus* has similar habits.

7 Creophilus maxillosus. A beetle very similar to the last species but often smaller and with a shinier head and thorax and a pattern of pale ashy-grey hairs on the wing-cases and hind body. The larger male has a broader head. The one British species is found on decaying animal matter, especially fish.

8 Mycetoporus splendidus. A quick-moving little beetle, often to be found in the winter. It does not live on fungi as its generic name suggests, but in moss, rotten wood, and damp dead leaves, and is fairly common over England, but rarer in the north.

9 Oxypoda alternans. A more or less gregarious little beetle to be found in decaying fungi all over the British Isles, usually in the autumn, and often with others of its very large sub-family.

10 Glow-worm (*Lampyris noctiluca*). The Glow-worm is not a Rove Beetle but more closely allied to the Soldier Beetles described on p. 174. Most members of this family are the tropical Fire-flies. They are remarkable for producing a soft, greenish light, brightest in the adult female, but existing in both sexes at all stages. The Glow-worm female is woodlouse-like, little different from the larvae except in size, and without wings or even wing-cases. It lives in places such as grassy banks, especially on chalk and limestone, but not, apparently, in Ireland. By day Glow-worms hide under stones or rubbish, or in the ground. By night the males fly, often in numbers, using large eyes to find their mates by the luminous organs in the underside of the female's tail segments. The female may climb up grasses or plants to show herself better. They feed on snails; both larvae and adults inject substances into the snail's body which first paralyse it, then make it dissolve. The Glow-worm then sucks up the liquid until the shell is empty.

TWICE LIFE SIZE (IN CIRCLES × 6)

1 OXYTELUS LAQUAETUS   2 OXYTELUS TETRACARINATUS   3 STENUS CICINDELOIDES

4 PAEDERUS LITTORALIS   5 PHILONTHUS POLITUS   6 OCYPUS OLENS

8 MYCETOPORUS SPLENDIDUS   7 CREOPHILUS MAXILLOSUS

9 OXYPODA ALTERNANS   10 LAMPYRIS NOCTILUCA + LARVA

SOLDIER AND CLICK BEETLES, ETC.

Nos. 1 – 6 on this page and the Glow-worm (p. 172) belong to a group of beetles with relatively soft, leathery body and wing-cases. The Soldier and Sailor Beetles (family Cantharidae) are long, narrow insects with long, robust legs and long, threadlike antennae. Most are active by day, seeking their insect prey on plants and flying readily. The larvae are dark, velvety, and flattish, with toothed jaws. They feed on earthworms and pupate in the soil in the spring.

Click Beetles or Skipjacks (family Elateridae) have hard body and wing-cases and are at least partly vegetarian. Their small heads are usually bent down, and they have short, thin legs and toothed antennae. When alarmed, they often slip to the ground and land on their backs. To right themselves, they bend their bodies backwards until they rest on their heads and tail tips; then they let go suddenly, springing upwards with a click. They may repeat this several times until they land the right way up. They will also skip, perhaps a considerable distance, to avoid an enemy. The larvae mostly resemble wireworms (No. 9). The larger species take several years to complete their growth. One group of tropical Fireflies are Click Beetles.

1-2 Soldier and Sailor Beetles. The Rustic Sailor Beetle (*Cantharis rustica*) (1) is found all over the country in early summer on nettles and grass, in fields and hedges. It is very like the Dark Sailor (*C. fusca*) which has the black spot on the front instead of the middle of the thorax and has all black legs. The Wood Sailor (*C. pellucida*) has a red thorax, abdomen, and legs, and is commonest in woods; and the smaller, similar Grey Sailor (*C. nigricans*) has greyish down on the wing-cases.

The Pale Soldier (*C. pallida*) (2) inhabits marshy localities, whereas the similar Common Soldier (*C. cryptica*), distinguished by the raised hairs on the wing-cases, is found everywhere. *C. livida* is like these but much larger. Very common on umbelliferous flowers such as hogweed, in late summer, is the Black-tipped Soldier (*Rhagonycha fulva*).

3 Yellow-tipped Malthine (*Malthinus flaveolus*). This is very like a small Soldier Beetle but has shortened wing-cases tipped with bright yellow, below which the ends of the folded wings show. It lives in and about woods, the little-known larvae inhabiting rotten wood. This is the largest of the group, most of which are small.

4 Red-tipped Flower Beetle (*Malachius bipustulatus*). A common beetle on flowers on the edges of fields and woods in early summer over most of England, though rare in Scotland and absent from Ireland. The larvae hunt their prey in decayed wood and under bark. When disturbed, the insect protrudes two pairs of scarlet, bladder-like organs, the function of which is not known.

5-6 Clerid Beetles (family Cleridae). Most of these are tropical insects, brightly coloured and often hairy. The hairy, usually pink or red larvae prey on other larvae. *Thanasimus formicarius* (6), the ant-beetle, is typical of the family and is widely distributed over Britain in wooded areas on or under the bark of dead trees. The Hide Beetles, such as *Necrobia ruficollis* (5), are less typical of the family. This not very common species is red and blue, but the more common *N. violacea* lacks the red. Either are to be found on dry

carcases, old bones, or skins, and they may fly to artificial light.

7 Mottled Skipjack (*Adelocera murina*). The covering of short, dense, brown and grey hairs give this broad, flattish beetle a mottled effect. It is commonest on bushes in early summer on sandy or limestone soils, or amongst grass or moss, and is rare in the north and Ireland. The larvae feed on roots.

8 Red-brown Skipjack (*Athous haemorrhoidalis*). A very common beetle everywhere in early summer, especially in woods on bracken. Other related species are *A. hirtus*, which is shining black and appears, much less frequently, later in the year, and the much bigger, rusty-brown *A. villosus*, which is most often seen as a black larva feeding on other larvae under the bark of dead trees.

9 Wireworm Beetle (*Agriotes obscurus*). This and the very similar *A. lineatus*, common round London, are best known for the destructive larvae which attack the roots of crops, especially in spring and autumn, and which may live in grassland for four or five years. The beetles appear early in the year and are nocturnal, spending the day under clods or at grass-roots.

10 Bordered Skipjack (*Dalopius marginatus*). A common beetle of trees and bushes in woods. The dark stripes down the yellow-brown wing-cases are sometimes almost absent. There are other groups of Skipjacks, some species passing at least their early stages in rotten wood. A large, golden-green species, *Corymbites pectinicornis*, is found in some grassy places in the Midlands and North.

11 Agrilus pannonicus. The Buprestid family, most species of which are large, metallic, tropical beetles, is allied to the Click Beetle family. The dozen British species are mostly small, and the nearly legless larvae are wood, stem, or leaf miners. *A. pannonicus* is now very rare and may be almost extinct in Britain. It used to be found in a wood in Kent, and later in the Sherwood Forest in Nottinghamshire.

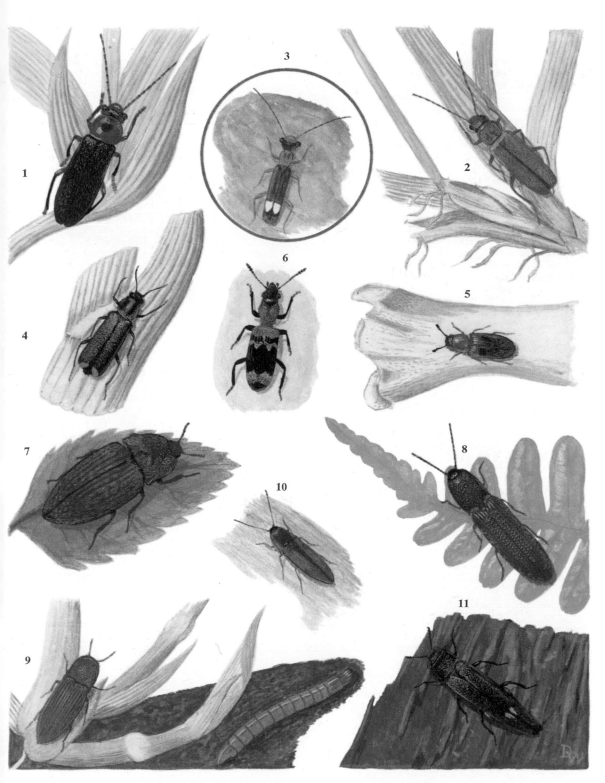

THREE TIMES LIFE SIZE (IN CIRCLE × 4)

1 CANTHARIS RUSTICA   3 MALTHINUS FLAVEOLUS   2 CANTHARIS PALLIDA
4 MALACHIUS BIPUSTULATUS   6 THANASIMUS FORMICARIUS   5 NECROBIA RUFICOLLIS
7 ADELOCERA MURINA   10 DALOPIUS MARGINATUS   8 ATHOUS HAEMORRHOIDALIS
        9 AGRIOTES OBSCURUS | LARVA   11 AGRILUS PANNONICUS

175

1-3 Skin Beetles (family Dermestidae). Some of these do great damage to preserved skins, furs, woollens, and animal products generally, especially when stored in warehouses or museums. It is the hairy, often tufted, jerkily-moving larvae which do most of the damage by feeding on the materials and, when ready to pupate, by boring into anything at hand, such as woodwork; they will even penetrate sheets of lead. The small or fairly small compact beetles usually fall to the ground when touched, or feign death. There are a dozen British species, but others have become established indoors.

The Bacon or Larder Beetle (*Dermestes lardarius*) (1) attacks most kinds of animal substances. It is found in houses, stores, beehives, and such places over most of Britain, and sometimes can be found outdoors in sheepskins. It once became such a serious fur-pest that the Hudson Bay Co. offered a reward of £200,000 for an effective remedy. It can be very troublesome in silkworm farms.

The Leather Beetle (*Dermestes maculatus*) (2), though plain above, has a thick snow-white pile underneath, varied by some black spots. It is often more common than the Bacon Beetle and equally widespread and destructive. In museums the larvae have been put to clean skulls and skeletons, which it is said they do better than any human. It hibernates in any stage, and produces up to six broods a year. Outdoors, Leather Beetles are found in carrion with their relatives the Mottled Skin Beetles (*D. murinus*).

The Museum Beetle (*Anthrenus museorum*) (3), a common beetle over England and Wales belonging to a group of small, round Skin Beetles which have scales rather than hairs; they have very slender legs and short antennae. Out of doors they are found on flowers, especially of parsley, in July. They pass the early stages in places such as birds' nests or among insect remains trapped in spiders' webs. They also live indoors, and the larvae may damage carpets, stored products, and insect collections, though the Common or Varied Carpet Beetle (*A. verbasci*) is a worse indoor pest.

4 Raspberry Beetle (*Byturus tomentosus*). A very common pest of raspberries, closely related to the Dermestidae, but with differently constructed feet and hairless, vegetarian larvae. After June, most specimens have lost their original golden-yellow gloss and appear worn and greenish-grey. The eggs are laid in raspberry and blackberry flowers, into which the young larvae bore and feed on the developing fruits. They pupate in the ground.

5 Ladybird Mimic (*Endomychus coccineus*). This striking beetle may be mistaken for a Ladybird, but it has very different antennae. The black patch on the thorax may be faint or absent. It lives gregariously with its larvae on or around fungoid growth under the bark of trees and stumps, and is fairly common locally over England and most of Scotland.

6- Ladybirds (family Coccinellidae). These well-known
11 and popular beetles, with their neat, bright, attractive appearance and lack of shyness, do much good by feeding, both as larvae and adults, on aphids (p. 34). Some of the 40 or more British species would hardly be recognized as ladybirds were it not for the apparently 3-jointed feet and very short, small-clubbed antennae. Their bright and boldly-contrasting colours warn predators such as birds that they are distasteful. The larvae are usually rather long-legged, pale-spotted, bluish-grey grubs which, like the adults, wander over the foliage of bushes and trees in search of their prey. The similarly-coloured pupae are attached by the tail to leaves. The beetles sometimes overwinter in large masses under loose bark, or indoors, in the corners of window-frames or such places. There may be several broods in the year.

The Red Marsh Ladybird (*Coccidula rufa*) (6) is different in shape from most Ladybirds and has no spots. It is found in marshy places, such as among reeds, all over England and Ireland, but not in northern Scotland.

The Two-spot Ladybird (*Adalia bipunctata*) (7) is perhaps the best-known and most abundant Ladybird, and is also about the most variable British insect in its colour-pattern. It usually has red wing-cases with two black spots, but many are black with some red spots. There is a whole range of different forms, but most, like the sandy yellow one shown here, are far less often seen. The underside and legs are always black. They are not always easy to distinguish from the almost equally common and variable Ten-spot Ladybird (*Adalia 10-punctata*) (8) of which the lower beetle shown in the picture is the commonest form. But the Ten-spot always has a brown underside with a small part of the breast whitish, and pale legs. Though often seen with the Two-spot Ladybird, it has a marked preference for trees and bushes in woods.

The Seven-spot Ladybird (*Coccinella 7-punctata*) (9) is much larger and hardly varies at all. It is also common over the whole of Britain, but not so abundant, except in certain years. It chiefly affects the lower-growing plants. A very similar species, the Ants'-nest Ladybird (*C. divaricata*) is found only about the nests of the wood ant.

The Twenty-two-spot Ladybird (*Thea 22-punctata*) (10) is small and neat, and varies very little. The beetle and its yellow, black-spotted larvae live among nettles, docks, grass, and low herbage, and are rather local, but not uncommon, over most of England, Wales, and eastern Ireland, but not in Scotland.

The largest British ladybird, the Eyed Ladybird (*Anatis ocellata*) (11), lives on pines, often with another large species, the Streaked Ladybird (*Neomysia oblongoguttata*), which has pale streaks instead of dark spots on the wing-cases.

THREE TIMES LIFE SIZE

1 Dermestes lardarius + larva 3 Anthrenus museorum 2 Dermestes maculatus
4 Byturus tomentosus 5 Endomychus coccineus
7 Adalia bipunctata 6 Coccidula rufa 8 Adalia 10-punctata 177
9 Coccinella 7-punctata 10 Thea 22-punctata 11 Anatis ocellata

Many unrelated beetles are woodborers, but here we refer only to the group Teredilia, which includes some of the most destructive beetles (*see* Nos. 1 – 6). Some species feed on fungi and stored products; none attack healthy trees; in fact, the most harmful are those which can infest already seasoned timber. These beetles, mostly small, usually have much-bent-down heads, often hidden by the hooded front of the thorax. Their larvae, shaped like tiny chafer grubs, tunnel in the solid wood and pupate in cells near the surface.

Nos. 7 – 9 on this page and 1 – 5 on p. 180 are Heteromera or 'Odd-toed' Beetles. They are varied in appearance and habits, but all have five-jointed front and middle feet but four-jointed hind feet. None are carnivorous as adults.

1   **Powder-post Beetle** (*Lyctus fuscus*). A small wood-borer, sometimes found in numbers on fresh oak palings or freshly-hewn surfaces of trees, around midsummer; it is a pest in stored timber, as is the similar but smoother *L. brunneus*. The beetles of this small family (Lyctidae) do not bend their heads down, have antennae with 2-jointed clubs, and feet with tiny first joints and long last joints.

2   **Death-watch Beetle** (*Xestobium rufovillosum*). This notorious and relatively large and stout woodborer can be found locally under the bark of old oaks and other trees, tunnelling in the decayed trunks, in the roof-timbers of churches, and woodwork in old houses, as far north as Lancashire. The beetle is sluggish and shuns light. At mating time the sexes call to each other by tapping their jaws on the wood — a sound which, heard in the stillness of the night, has given rise to the superstition connected with their name.

3   **Furniture Beetle** (*Anobium punctatum*). A common household pest, the larvae of which often attack furniture and other woodwork and, if unchecked, will reduce wood to a shell full of powder; small round holes in woodwork are usually a sign of its presence. It is also found under the bark of old trees. The adults gnaw their way out and fly at night in June and July. They feign death when disturbed.

4   **Fan-bearing Woodborer** (*Ptilinus pectinicornis*). The male has beautifully fan-like antennae, which in the female are merely saw-toothed. It is found in June and July, on trees, especially beech, over most of Englnad and Wales, but is rare elsewhere. It runs quite actively on dead trunks, and flies in the sunshine. The escaping beetles make groups of neat, round holes in wood stripped of bark.

5-6 **Spider Beetles** (family Ptinidae). Although these long-legged, often round-bodied, active, spider-like beetles belong to the woodboring group, many have become adapted to other ways of life. Both the species shown here feed on dry animal and vegetable matter, in houses and stores over most of the British Isles, but are now less common than the Common Spider Beetle (*Ptinus tectus*). The male White-marked Spider Beetle (*Ptinus fur*) is narrower than the female (5), with straight sides and very long legs. In fact, the white marks are often very slight, and are much more obvious in the Six-spotted Spider Beetle (*P. sexpunctatus*), often found in houses. The sexes are alike in the Golden Spider Beetle (*Niptus hololeucus*) (6).

7   **Thick-legged Flower Beetle** (*Oedemera nobilis*). The male has thickened hind legs, those of the female being quite normal. This bright beetle frequents flowers in May and June in southern England, chiefly towards the west, and flies actively in the sunshine. The larvae feed in plant stems, but some species develop in dead wood.

8   **Cardinal Beetle** (*Pyrochroa serraticornis*). A handsome beetle found in June on flowers and herbage such as nettles, along hedgerows and wood-margins, from the Midlands southwards. The Black-headed Cardinal (*P. coccinea*), is larger, a deeper blood-red with a black head, and frequents old woodland and park-land. The yellowish larvae of both beetles are long and flattened, with two short prongs at the tail end; they live under the bark of fallen trees and stumps.

9- **Oil Beetles** (family Meloidae). These beetles all have a
10 special chemical in their blood. With the Blister Beetle or Spanish Fly (*Lytta vesicatoria*) (9) this chemical was used medicinally as a blistering agent, the bodies of the beetle being dried and pounded up to obtain the powerful irritant, cantharidin. The Oil Beetle (*Meloë proscarabaeus*) (10) exudes an evil-smelling fluid from its joints to deter enemies, and its wing-cases overlap one another, which is very unusual in beetles.

The Blister Beetle feeds on ash, privet, and lilac in July, and is rare, except occasionally in places in southern and eastern England. The Oil Beetle has a remarkable life-history. The female becomes greatly swollen and lays thousands of tiny eggs in batches in cracks or holes in the ground; these hatch into long-legged little larvae which swarm over plants in hot weather. Only those survive which manage to attach themselves to a wild bee and are carried to the bee's nest. There, they enter brood-cells and at once devour the egg or young grub in the cell. They change into chafer-like grubs, feed on the store of honey, and grow fast. They pass the winter as headless, legless maggots, then change shape again, rest for a time, and finally pupate. The wingless adults crawl about in the spring sunshine on heaths and open grassland nibbling at buttercup plants. Of the seven British species only this and the Bluish Oil Beetle (*M. violaceus*) are at all common; and even they are now quite local.

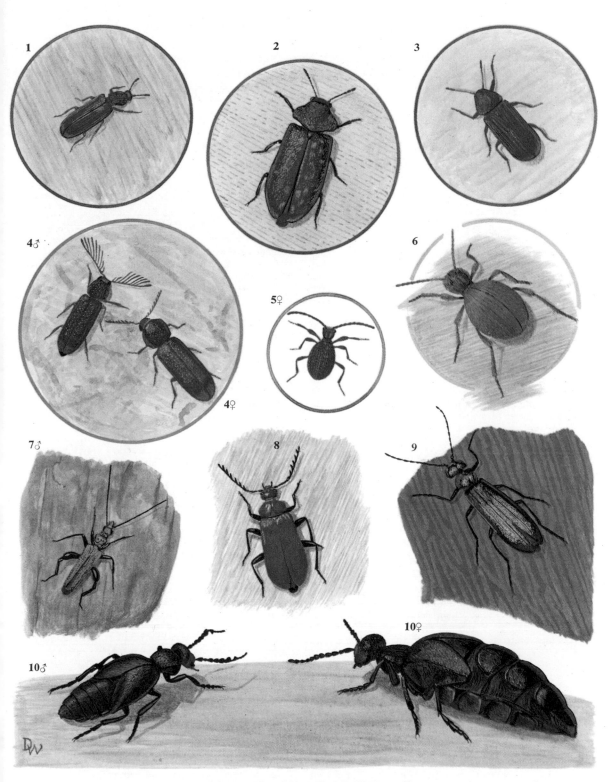

TWICE LIFE SIZE (IN CIRCLES × 5)

1 Lyctus fuscus	2 Xestobium rufovillosum	3 Anobium punctatum
4 Ptilinus pectinicornis	5 Ptinus fur	6 Niptus hololeucus
7 Oedemera nobilis	8 Pyrochroa serraticornis	9 Lytta vesicatoria
	10 Meloë proscarabaeus	

179

The 'False Ground Beetles' (Tenebrionidae, Nos. 1 – 5), belong to the Heteromera group, about which there is a note on p. 178. They are mostly ground-living, often on sand-dunes, but some live in tree-fungi or under bark, and others in stored grain. The wireworm-like larvae are nearly all vegetarian.

The Dung Beetles (6 – 10) are closely allied to the Chafers (p. 182) with the same robust, thick-set build and similar structure, especially of the antennae. The larvae, too, are similar. The beetles usually have very powerful, spiny legs adapted for digging, but the almost useless feet are mostly much shorter and thinner than in the Chafers. The larger ones can make a high, sharp chirping sound by moving the abdomen against the wing-cases; these fly in the late evening and night, whereas the smaller species fly mostly by day. They all have a one-year life-cycle, and a few feed on fungi or rotting vegetable matter instead of dung.

1 **Churchyard Beetle** (*Blaps mucronata*). A large, ill-smelling, wingless insect, sometimes called the Cellar Beetle, which stalks about at night in dark places such as cellars and crypts, and even in the deepest coal mines, but is rarely seen in open country. It is quite common and long-lived.

2 **Broad-horned Flour Beetle** (*Gnathocerus cornutus*). The male of this small beetle has a curiously-shaped head, with a broad, ear-like 'horn' at each side, a pair of small horns or teeth between, and lengthened and upward-curved jaws. It spends its life in old flour and is sometimes found baked into bread.

3 **Mealworm Beetle** (*Tenebrio molitor*). The larvae are the well-known mealworms, pests of flour-mills and corn stores, and bred on a large scale as a food for caged animals and birds. All stages live together in flour or bran, where there may be several broods a year. Outdoors there is only one generation, and they breed in pigeons' nests, chicken coops, and such places. The beetles fly at dusk and after dark, often being attracted to lights in July and August.

4 **Blue Helops** (*Helops caeruleus*). A handsome beetle varying from greenish-blue to almost purple above. The female is larger than the male. It lives by day under loose bark and in rotten wood, but at dusk it emerges into the open and runs about on the trunks. Its cylindrical, pale-coloured larva has a pair of sharp upwardly-curved tail-spines. It is found scattered about southern England, but only very locally.

5 **Common Helops** (*Cylindronotus laevioctostriatus*). A similar but smaller beetle with shorter legs and body. It is common over most of England and Ireland, sometimes being found in clusters under pine bark and also under heather and moss, or at tree roots. It is known to feed on tree-algae. Both it and the Blue Helops emit a peculiar smell like that of the Churchyard Beetle.

6 **English Scarab** (*Copris lunaris*). A relative of the famous Sacred Scarab worshipped by the ancient Egyptians, but now very rare, being last found in any numbers in Surrey in 1948. It makes deep, slanting burrows under cow or horse dung, usually in sandy ground, at the end of which the female hollows out a large chamber. She provisions this with balls of dung, laying an egg in each ball; she then mounts guard over them until the young are safely established.

7 **Minotaur** (*Typhaeus typhoeus*). In spite of the male's horns, this dung beetle is closely related not to the Scarab, but to the Dor Beetles. from which the horn-less female is distinguished by having no trace of metallic colouring. Minotaurs are locally common, mainly in spring and autumn, on sandy heaths as far north as Ayrshire, and are very local in Ireland. Single ones are often seen crawling on pathways or amongst grass. They dig deep burrows, usually under cow or rabbit dung, in which both sexes stack rolls of dung, filling most of the burrow; an egg is laid in each roll.

8 **Common Dor Beetle** (*Geotrupes stercorarius*). The word 'dor' comes from an old word meaning 'drone', for in the evening this beetle has a humming flight, after which it often comes to a light. Its hairy under-side is a bright metallic blue, green, violet, or copper. Both sexes burrow under cattle or horse dung and accumulate a store of dung at the end of the burrow, in which the eggs are laid. This and the very similar *G. spiniger* are common throughout Britain. The smaller, shorter Wood Dor Beetle (*G. stercorosus*) has less deeply grooved and more wrinkled wing-cases, and is often common in wooded places. They are all active up to very late in the year, and overwinter in their burrows.

9- **Dung Beetles** (genus *Aphodius*). There are over 40
10 species, all oblong and relatively small. Instead of digging burrows, they lay eggs in the dung where it is, though the larvae pupate in the ground. Both *A. fimetarius* (9) and *A. rufipes* (10) use the dung of horses, cattle, and sheep, and are common everywhere. *A. rufipes* is the largest of the genus, and the only one to fly habitually at night.

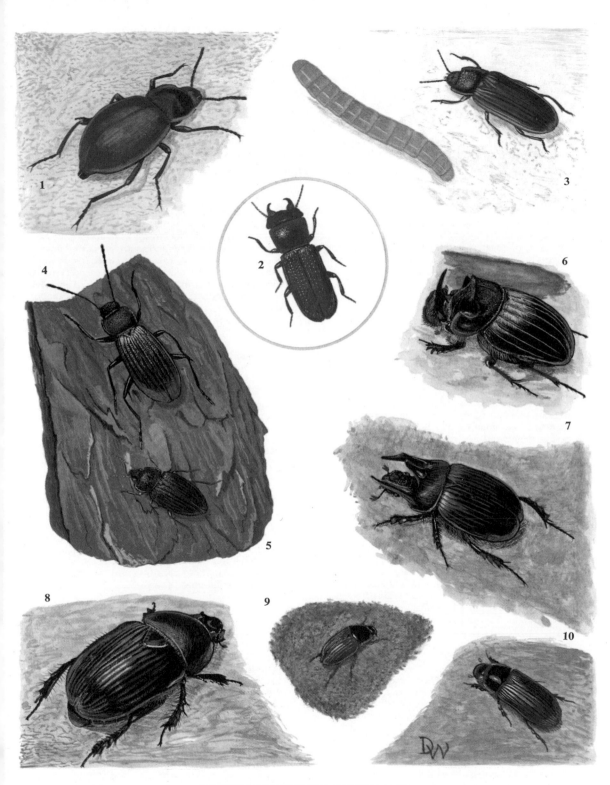

TWICE LIFE SIZE (IN CIRCLE × 8)

1 Blaps mucronata   2 Gnathocerus cornutus   3 Tenebrio molitor + larva

4 Helops caeruleus   6 Copris lunaris

5 Cylindronotus laevioctostriatus   7 Typhaeus typhoeus

8 Geotrupes stercorarius   9 Aphodius fimetarius   10 Aphodius rufipes

These and the Dung Beetles (p. 181) form a large section called Leaf-horn Beetles (Lamellicornia). They have short antennae with clubs formed of thin movable plates (leaves) which spread out fanwise when the insects are active. The antennae of Stag Beetles (6 – 7) are longer, elbowed, and with fixed leaves to the clubs. Chafers (the word means 'biter') belong to the Scarabaeidae, which includes the biggest known beetles and is very numerous in the tropics. The thick, curved larvae feed on some form of vegetable matter, and the root-feeding ones pupate deep in the ground. The beetles often feed on the nectar of flowers, and sometimes eat foliage, but adult stag beetles take little or no food. They all spend the winter as adults, but still in the pupal cell.

1   Summer Chafer (*Amphimallon solstitialis*). A smaller, hairier relative of the Cockchafer, local but widespread in England and Wales. The larvae live in turf, which they may damage. The beetles fly back and forth, usually in companies, on warm July evenings before dusk, usually keeping to a particular 'beat' — round a group of trees, or in one corner of a field, for example. A very similar but rare species, the Scarce Summer Chafer (*A. ochraceus*), is a day flier.

2   Garden Chafer (*Phyllopertha horticola*). This beetle, often called the Field Chafer or June Bug, varies a good deal, the fore body and underside being sometimes a bright green, the legs most commonly being quite dark, and the wing-cases often darker down the middle. The feet end in a pair of unequal, strong, hooked claws. It used to be common all over Britain, and was often a pest in fields and gardens, but it is now only at all plentiful in certain years, mostly in woods on bushes, bracken, or flowers, and here and there on the slopes of downs, flying in June in sunshine. Anglers use the beetles for bait, giving them many names.

3   Bee Chafer (*Trichius fasciatus*). A relative of the Rose Chafer, so called because, except for the wing-cases, it is covered with shaggy, yellowish hair, giving it a bee-like aspect when in flight, and so presumably some protection from predators. It is not uncommon in parts of the Scottish Highlands and Wales, frequenting flowers in the June and July sunshine. The early stages are passed in the wood-mould of old birch stumps.

4   Cockchafer (*Melolontha melolontha*). This is also known as the 'May Bug'. When fresh, the wing-cases are thickly powdered with white. The male has long fan-like clubs on the antennae. It rests by day in trees and flies in the evening and at night, visiting flowers and often entering lighted rooms. It occasionally occurs in vast swarms, when the larvae, which live at least three years, do great damage to crops and grassland, and the beetles strip trees of their leaves. On one occasion, 80 bushels of them were collected on one farm. Fortunately such outbreaks are now very rare.

5   Rose Chafer (*Cetonia aurata*). This beautiful insect, with a golden gloss over its shining green, sometimes coppery, upperside, flies readily. It appears in early summer in the midlands and south, on various flowers, and is said to damage roses. The larvae live in rotten stumps, leaf-mould, old compost-heaps, and even, like the much duller Northern Rose Chafer (*C. cuprea*), in wood-ants' nests. *C. aurata* is now more local and less plentiful than it used to be. This group of chafers includes many gorgeously-coloured tropical species.

6   Stag Beetle (*Lucanus cervus*). The males are the giants among British beetles, the largest being not far short of three inches in length, part of which is accounted for by the huge antler-like jaws. Their size, however, varies widely, and the smallest have jaws not much longer than the head, with less pronounced teeth. The smaller female is very different and varies less in size. The male, when accosted, rears himself up by means of his fore legs, raises his head, and opens his jaws wide; but this menacing attitude is largely bluff, and the fearsome-looking jaws cannot, in fact, inflict such a painful bite as those of many smaller beetles. The adults can sometimes be dug out of old stumps, but are more often to be seen, from May to July, on tree-trunks, fences, and pavements, and are not uncommon in the London suburbs. They fly from dusk onwards on sultry nights and may come to artificial light. The big larvae live for several years in the roots of rotting stumps. The species is locally spread over the southern half of England only.

7   Lesser Stag Beetle (*Dorcus parallelipipedus*). The sexes differ far less than with the Stag Beetle, but the male's fore parts are broader and very dull, and the female's narrower (especially the head) and shinier. It lives, with its larvae, in rotten wood, mainly of beech, elm, and ash; and being long-lived, can be found throughout the year. It is nocturnal, sluggish, and not very often seen in the open, and is common in many parts of England, south of Nottingham or thereabouts.
The Rhinoceros Beetle (*Sinodendron cylindricum*) has similar habits, and is more widespread but in some parts more local. It looks little like one of this family, being cylindrical, and even the male has a small head and jaws; the male also has a curved horn on the head. The beetle is very shiny black and coarsely sculptured.

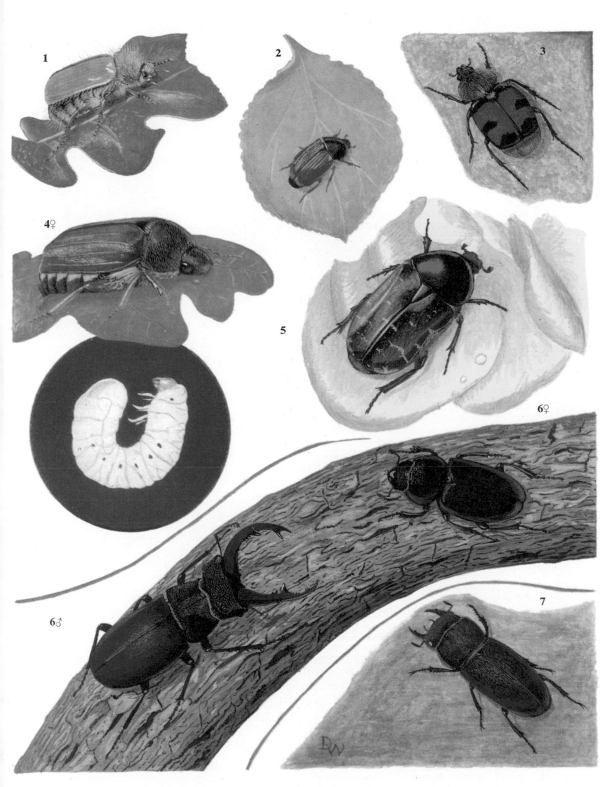

TWICE LIFE SIZE (IN CIRCLE AND No. 6 LIFE SIZE)

1 Amphimallon solstitialis   2 Phyllopertha horticola   3 Trichius fasciatus
4 Melolontha melolontha + larva   5 Cetonia aurata
6 Lucanus cervus   7 Dorcus parallelipipedus

183

LONGHORN BEETLES (*Cerambycidae*)

About 60 species of Longhorn (Longicorn) Beetles are British, though others are sometimes introduced in timber from abroad. There are endless species in the great tropical forests — often elegantly shaped and handsomely coloured. Most Longhorns are long and narrow, with long legs, and, especially the males, with long or very long antennae. They have short, powerful jaws, directed downwards. Most, when disturbed, make a creaking noise by rubbing the thorax against the base of the hind body. The adults if they eat at all eat only nectar. The larvae are fleshy, wrinkled, whitish grubs, with horny heads and short or no legs. Most pupate in the wood in which they feed, often taking several years to mature.

1 **Tanner Beetle** (*Prionus coriarius*). The biggest British Longhorn, the females being often over 1½ inches long, is unique for Britain in having three spines on each side of the thorax and saw-toothed antennae, very stout in the male. It is the only British member of its group, which includes some very large beetles. Tanners are found south of Lancashire in old park or forest land in late summer; they fly at dusk or after, sometimes in small swarms, and are sluggish by day, resting on tree-trunks or stumps. The larvae feed in the roots of dying or dead trees, and pupate in large underground cells by the roots, where the adults sometimes hibernate. The males are said to fight each other fiercely.

2 **Two-banded Longhorn** (*Rhagium bifasciatum*). A common insect over most of Britain, usually found in rotting pine stumps or posts. Its pale markings vary a good deal. Also common is the Eyed Longhorn (*R. mordax*), a yellow beetle with two eye-like black spots. The larvae mine galleries under the bark of fallen trees, mainly oaks.

3 **Variable Longhorn** (*Stenocorus meridianus*). The wing-cases are sometimes all or partly black; and the legs and antennae either black or red. The size also varies, large females being several times as large as small males. They fly readily and frequent flowers, especially at the edges of woods, from May to July, over all England, though seldom in number, and less frequently in the north. The larvae live in the hard wood of stumps or dead boughs.

4 **Spotted Longhorn** (*Strangalia maculata*). Though varying less in size than the last, it varies greatly in the extent of the black markings. It is found fairly commonly from June to early August on flowers in or about woods in most parts of England and Ireland. The larvae feed in stumps and dead boughs of birch and other trees, as do those of the less common, often larger Four-banded Longhorn (*S. 4-fasciata*).

5 **Musk Beetle** (*Aromia moschata*). Its pleasant musk-like scent gives this beetle its name. Its metallic colouring may be gold-green, dark copper, blue-green, or purple-black. The male has longer antennae than the female — longer than its body. It lives on osiers, willows, and old sallows, in which the larvae burrow, sometimes doing severe damage. From June to September the beetles occasionally sun themselves in numbers on the boles of the food trees, and may visit nearby flowers. They used to be common, but in the south at least are now far more local.

6 **House Longhorn** (*Hylotrupes bajulus*). This tiresome, sometimes serious, indoor pest is fortunately rare, and very rare outdoors. It has two raised, shining spots on the white-haired thorax, and a slight band of white hair across the wing-cases. It is found mostly in parts of Surrey, but not in large numbers since about 1950–53. The larvae tunnel in dry pinewood, for example in roof- and floor-timbers, sometimes for as long as 12 years. The nocturnal adults appear in late summer and after breeding soon die.

7 **Wasp Beetle** (*Clytus arietis*). The quick, jerky movements as well as the colours of this beetle make it resemble some solitary wasps — an example of 'protective mimicry' (*see* p. 155). It more often basks on sunlit foliage than on flowers; and is fairly common over England in June, especially in the south. The larvae bore into dead wood such as posts or fallen trunks.

8 **Weaver Beetle** (*Lamia textor*). A very rare, massively-built Longhorn, with thick feet and base of the tapering antennae, which lives on willows, osiers, and sallows. It is so named because it is supposed to weave together twigs or shoots as a nest. The larvae feed for two years in the roots or base of the trunk, and the sluggish nocturnal adult creeps on the ground or low on the tree from July to September.

9 **Timberman** (*Acanthocinus aedilis*). The male's antennae are four times as long as its body; those of the female twice as long. It breeds only in the older pine forests of the Scottish Highlands; but odd specimens are transported elsewhere in pine logs. The flattened, legless larvae dig galleries under the bark of fallen or felled pines, where they pupate about midsummer, the long antennae coiled round the body. The beetles hibernate fully formed in the pupal cells and emerge the next May or June, when they fly in the sun with their antennae trailing behind them.

10 **Poplar Longhorn** (*Saperda carcharias*). A large beetle, usually found on or about old poplars from July to September. The larvae feed from two to three years in trunks or branches, and the adults eat holes in the leaves. It is rare in the south but found through the eastern counties to the north of Scotland, in some years being fairly common in East Anglia.

LIFE SIZE

1 PRIONUS CORIARIUS   2 RHAGIUM BIFASCIATUM   3 STENOCORUS MERIDIANUS

4 STRANGALIA MACULATA   5 AROMIA MOSCHATA   6 HYLOTRUPES BAJALUS

7 CLYTUS ARIETIS   8 LAMIA TEXTOR

9 ACANTHOCINUS AEDILIS   10 SAPERDA CARCHARIAS

LEAF AND FLEA BEETLES (*Chrysomelidae*)

The majority of this family, which includes the Tortoise Beetles (p. 188), are much rounded, smooth, and often with metallic colouring. Typically they have small heads sunk in the thorax and rather short legs and antennae, and all have four-jointed feet. Most enjoy sunshine, and if disturbed, usually draw in their limbs and fall to the ground, or, in the case of Flea Beetles, jump. The larvae are shaped like woodlice or are caterpillar-like; those which feed openly are coloured, as are those pupae which are not hidden.

1-3 Reed Beetles (tribe *Donaciini*). These long, narrow beetles are shaped rather like Longhorns (p. 184), with lovely metallic tints and their undersides covered with a dense, short silky pile which repels water. They live on water plants, especially reeds, and fly very readily. The larvae are aquatic and obtain air by inserting tail-spiracles into the air-spaces in the stems of water plants. They pupate in cocoons attached to underwater stems or roots.

The Red and Green Reed Beetle (*Donacia aquatica*) (1) is widely distributed, found especially on floating reed-grass, but like several of the species, it is very local. It has long hind thighs with a small tooth on the inside, and is golden coloured beneath. The Striped Reed Beetle (*D. vulgaris*) (2) is silvery beneath and has shorter hind thighs with no tooth and an often quite bright broken stripe along the back. It is found on reed-mace and sometimes bur-reed, and is much less common than the Plain Reed Beetle (*D. simplex*), which is similar but with no stripe. The Variable Reed Beetle (*Plateumaris sericea*) (3) is usually a shining bronze or green, but may be red, blue, or black, and the hind thighs are short and strongly toothed. It is common on various water plants.

4 Two-coloured Leaf Beetle (*Chrysolina*, or *Chrysomela*, *polita*). Common beetles over much of Britain on mint and other low plants. They have gold-green undersides, thorax, and legs. They are sluggish and shy, often lurking under cover by day; they hibernate as adults. *C. sanguinolenta* is blue-black with red edges; *C. banksi* is large and bright bronze; *C. violacea*, found in chalky places, is violet with red feet.

5 Colorado Beetle (*Leptinotarsa* 10-*lineata*). This very serious pest of potato fields is a native of the Rocky Mountains, U.S.A. Anyone finding it in any stage must notify the Ministry of Agriculture at once, for this is the only way of checking possible outbreaks. The adults and larvae, both of which eat the leaves, are so striking in appearance that they are easy to recognize.

6 Dock Leaf Beetle (*Gastrophysa viridula*). This metallic green beetle is found locally throughout Britain on dock. It is less common than its smaller relative *G. polygoni*, which has a red thorax and partly red legs, and is found also on knotgrass. The females of this genus may become so swollen with eggs that the wing-cases are forced upwards and apart.

7 Watercress Beetle (*Phaedon cochleariae*). A short, metallic-blue beetle which lives with its dark larvae on various kinds of cress, or sometimes on mustard crops, being also called the Mustard Beetle. It is common over most of Britain.

8 Poplar Leaf Beetle (*Chrysomela*, or *Melasoma*, *populi*). The bright-red colour becomes duller and browner in preserved specimens. It lives, often with the very similar *C. tremulae*, on aspens in woods, and also on poplars and sallows. There may be several broods in a year, producing large numbers for a few seasons, and then almost disappearing. The larvae, when touched, give off an oppressive tarry smell from glands opening along the sides. Two related species are spotted like Ladybirds.

9 Bloody-nosed Beetle (*Timarcha tenebricosa*). This largest of the British Leaf Beetles, when handled, exudes a red liquid from its mouth. Its body is usually more globular than shown here. The smaller and shinier male has very broad feet. The beetle feeds on bedstraw on heaths, roadsides, or hedge banks and is often seen crawling lazily amongst grass. The green-black, wrinkled larva is shaped much like the adult. Though considered common in the south of England, it now seems to be very local except towards the west. On similar plants lives the Bedstraw Beetle (*Sermyla halensis*), yellowish with the crown of the head and the wing-cases brilliant blue-green, and often common in late summer on downs, cliffs, or sandy tracts by the sea.

10-Flea Beetles (tribe *Halticini*). These very small, very
13 numerous beetles jump with their hind legs, which have noticeably thickened thighs. There are about 130 British species, some of which do great damage to crops. The larvae of most small Flea Beetles feed under the outer 'skin' of the leaves, where they pupate, and under which the eggs are inserted. There may be several broods in a year.

The Turnip Flea Beetle (*Phyllotreta nemorum*) (10) lives on plants of the cabbage family. This, and two very similar species, all called 'turnip flea' or 'turnip fly', are more or less common and often destructive. The Sallow Flea Beetle (*Chalcoides fulvicornis*) (11), locally common on sallows, is bright green or sometimes coppery or dark blue; *C. aurata*, found on willows and poplars, is vivid green with crimson fore parts, and the larger, rounder, aspen-feeding *C. aurea* is golden-copper, green, or blue. The Mangold Flea Beetle (*Chaetocnema concinna*) (12), common everywhere amongst grass and low herbage, is a bright brassy colour, with regular rows of deep punctures on the wing-cases. The Potato Flea Beetle (*Psylliodes affinis*) (13), very common on woody nightshade, also attacks related plants such as potato.

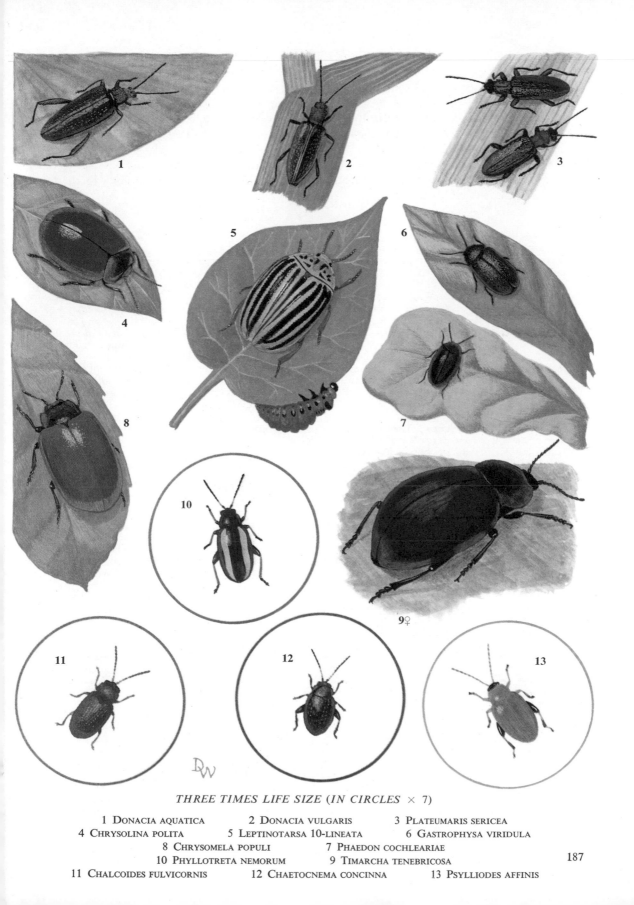

THREE TIMES LIFE SIZE (IN CIRCLES × 7)

1 Donacia aquatica	2 Donacia vulgaris	3 Plateumaris sericea
4 Chrysolina polita	5 Leptinotarsa 10-lineata	6 Gastrophysa viridula
8 Chrysomela populi	7 Phaedon cochleariae	
10 Phyllotreta nemorum	9 Timarcha tenebricosa	
11 Chalcoides fulvicornis	12 Chaetocnema concinna	13 Psylliodes affinis

187

TORTOISE BEETLES AND WEEVILS

Weevils are plant-feeding beetles, most with four-jointed feet and heads drawn out into a sort of snout (rostrum), which varies in length and shape between species. When long and thin, the female's being the longer, it is usually curved downwards. The antennae are usually clubbed and elbowed at the end of the long first joint to enable the insect, when at rest, to fold them back close to the snout. There are some 500 British species, mostly small. The usually whitish, legless, stout larvae live at or inside some part of the foodplant. Many overwinter as adults; few fly much, and many are wingless and extremely sluggish. Typical weevils (Nos. 7 – 12) belong to the family Curculionidae (*see* also p. 190).

1-2 Tortoise Beetles (sub-family Cassidinae). These curious Leaf Beetles (*see* p. 186) have thorax and wing-cases flattened into a sort of shield, completely covering the head and much of the legs. The broad, flat larvae have bristly spines at the sides, and a sort of recurved fork above the tail, upon which excrement collects and finally forms a covering both to hide the larva and to protect it from the sun. The flattish, spined pupa is attached to a leaf. The bright-green Common Tortoise Beetle (*Cassida rubiginosa*) (1), is found almost anywhere on thistles in summer. The Striped Tortoise Beetle (*C. nobilis*) (2), found at roots of plants of the pink family, is a greenish straw-colour, with a silvery-bluish stripe down each side of the back, which soon vanishes in dried specimens.

3 Pea Weevil (*Bruchus pisorum*). This, in fact, is not a weevil, but is more closely related to Leaf Beetles. It is hardly ever found outdoors, but is a pest of dried peas, in which all stages are passed. The rather similar *B. rufimanus* is found in the open, for instance in bean fields, and sometimes on flowers.

4 Red and Bronze Weevil (*Rhynchites aequatus*). A small, richly-coloured weevil, the female being larger with a longer snout. It is locally common on hawthorn flowers in southern England. The beetles lay their eggs in freshly-set fruits, and the larvae feed in the berries and pupate underground. In this group the antennae are not elbowed.

5 Red Oak Roller (*Attelabus nitens*). A smooth, hairless weevil, the male having long forelegs. In early summer the female lays her eggs on a young oak leaf, doubles the leaf over along the midrib, and rolls it into a short case as food for the larvae. She then bites half through the leaf-stalk, and consequently the case falls when the larvae are ready to pupate. The Red Hazel Roller (*Apoderus coryli*) does much the same thing with hazel leaves. Both are local but widespread.

6 Pear-shaped Clover Weevil (*Apion pisi*). This belongs to a large genus of small weevils, now considered a separate family. Many live on plants of the pea family, the larvae feeding on the seeds and pupating in the pods. *A. pisi*, whose deep-blue wing-cases are very broad behind, is fairly common all over Britain.

7 Coal-black Weevil (*Otiorrhynchus clavipes*). A large wingless weevil, varying in shape and size. When quite fresh, the wing-cases have scattered patches of pale hairs, and the legs are usually reddish. It occurs in moss and at roots of plants, chiefly in chalk or limestone districts in southern England, and in early summer also on hedges, fruit trees, and shrubs. With some related species the females lay fertile eggs without having mated, males being almost unknown.

8 Light-green Tree Weevil (*Phyllobius calcaratus*). The coating of fine, hair-like, green scales, slightly golden-tinted when fresh, later becomes worn and patchy. The broader female may be browner than the male. There is a strong tooth on each thigh, and the legs are reddish in contrast to the Green Nettle Weevil (*P. urticae*). The smaller Gold-green Tree Weevil (*P. argentatus*) is covered with lustrous, round scales. All these beetles appear in May or June; the early stages are passed in the ground.

9 Speckled Tree Weevil (*Polydrosus cervinus*). A fawn or palish-green beetle, covered with dark specks and patches, to be seen commonly in June on young trees in woods and hedges.

10 Water-plantain Weevil (*Hydronomus alismatis*). One of a group of small, sluggish weevils associated with water plants, and hard to find and identify. This one is locally common on water plantains over most of the British Isles. The larvae mine the leaves.

11 Apple-blossom Weevil (*Anthonomus pomorum*). A pest of apple trees wherever it occurs, this beetle emerges in autumn, hibernates in cracks or under bark, and reappears in spring to lay its eggs in flower buds — one to a bud. The larva eats all the flower except the petals, which form a sort of case round it in which it pupates in the summer. This and most others of the genus have long forelegs with strongly-toothed thighs, and a band of whitish hairs on the back.

12 Nut Weevil (*Curculio nucum*). As soon as the young fruit has set on hazel bushes, especially in southern England, the female Nut Weevil bores a hole into the kernel with her very long snout and there lays a single egg. When the grub hatches, it feeds until the nut falls in the autumn; then it eats its way out and burrows into the earth. The beetle appears in spring when the hazel comes into leaf. Other species attack acorns; and all have a similar triangular shape, variegated by scale-like hairs.

FOUR TIMES LIFE SIZE

1 Cassida rubiginosa   2 Cassida nobilis   4 Rhynchites aequatus

3 Bruchus pisorum + larva

5 Attelabus nitens   6 Apion pisi   7 Otiorrhynchus clavipes

8 Phyllobius calcaratus   9 Polydrosus cervinus

10 Hydronomus alismatis   11 Anthonomus pomorum   12 Curculio nucum

There is a note about weevils (Nos. 1 – 8) on p. 188.

1 Large Pine Weevil (*Hylobius abietis*). This pest of young pines, and occasionally of spruce or larch, feeds on the tender bark of shoots, sometimes almost stripping and so killing them; it also eats the buds. The grubs bore into old stumps and roots, doing far less damage. It is less common than it was in England, but is plentiful in Scotland.

2 Sandy Clover Weevil (*Hypera punctata*). This is the largest of the 19 British species, a group whose green larvae feed mostly on the pea family, and pupate in open network cocoons. It lurks by day under moss or rubbish, or at roots, feeding, either low down or at night, on clover; it sometimes crawls about in the open or even flies in hot weather. It is not uncommon.

3 Grain Weevil (*Sitophilus granarius*). A small but troublesome pest in granaries and corn stores, but is now controlled, mainly by fumigation. The female bores a hole in a grain of cereal and lays an egg in it; the larva gradually hollows out the grain and pupates in the husk. A single pair can give rise to over 6,000 beetles in a season, but cannot breed out of doors in Britain. The Rice Weevil (*S. oryzae*) attacks rice in the same way. The big tropical Palm Weevils, whose larvae the natives eat, attack coconut palms.

4 Osier Weevil (*Cryptorhynchus lapathi*). This stoutly-built weevil rests with its snout tightly folded in under its thorax, its legs folded up, and its knees sticking out. It is often very variegated, but always has a large pale patch at the end of the wing-cases: it is roughly scaled and set with raised black tufts. The larvae bore winding galleries in the stems of willows or sallows. It is rather a local and elusive beetle, rare in Scotland and Ireland.

5 Common Nettle Weevil (*Cidnorhinus 4-maculatus*). A member (as is No. 6) of a large group of small, short, squarish weevils, often with a tooth at each side of the thorax, patterns of light scales, and curved snouts, usually tucked under the thorax. If disturbed, they feign death and can be mistaken for seeds or bits of earth. This one is very common on nettles, the larvae feeding in the roots or base of stems.

6 Turnip Weevil (*Ceuthorhynchus pleurostigma*). Often called the Cabbage Gall Weevil, this belongs to the same group as No. 5. Its underside is whitish-scaled. It is common on cabbages, turnips, and related wild plants. The larvae live and feed in the galls they cause to form on the roots; they pupate in the soil. The adults emerge in late summer, overwinter, and reappear in the spring.

7 Figwort Weevil (*Cionus hortulanus*). A member of a genus of small, variegated weevils with narrow thorax and broad shoulders, living on figwort or mullein. The slug-like little larvae, covered with sticky slime, feed openly on the leaves, and pupate attached to the plant, in smooth, roundish cocoons made from the hardened slime. The rather smaller *C. hortulanus* is less black, and the yet smaller *C. alauda* is mostly white. All are locally common, often feeding together.

8 Elm Hopper Weevil (*Rhynchaenus saltator*). One of about 15 British weevils which have much thickened hind thighs and jump like Flea Beetles (p. 186). The snout is bent down or tucked right underneath, and the eyes nearly touch. The larvae mine the leaves of trees and bushes, pupating in the mines. It is often common on elm in the midlands and south.

9- Bark Beetles (family Scolytidae). These are weevils
11 adapted for burrowing in wood. They have very short or no snouts, and in many species the head is bent under the thorax, the feet are thread-like, and the body is set with fine upright hairs. Both adults and larvae tunnel galleries under the bark of fallen or dying trees and branches, the adults doing the most damage. The Ambrosia Beetles (No. 11) bore into solid wood, feeding on fungi to be found growing in the tunnels. Most Bark Beetles can be found throughout the year. The Large Elm-bark Beetle (*Scolytus scolytus*) (9) and the similar but smaller *S. multistriatus* are common in the south, flying freely in April or May, and a second generation in July or August. The larvae make radiating galleries under elm bark, doing damage to trees in parks and avenues. They spread a fungus infection, Dutch Elm disease.

The Black Pine Beetle (*Hylastes ater*) (10) is a serious pest all over England and Scotland in pine plantations, killing young trees by destroying the bark of both roots and below-ground stems. It is less often found in the thicker bark of trunks and logs. The equally destructive, broader, shinier *Blastophagus piniperda* first attacks the shoots and then the bark of less vigorous pines, but in large numbers can kill even healthy trees. The Shot-hole Borer (*Anisandrus dispar*) (11), one of the Ambrosia Beetles, is found locally in southern England. The wingless male, which never appears in the open and is seldom seen, is much shorter, rounder, and flatter than the female. This beetle bores into the solid wood of oak stumps or of holly, plum, and other fruit trees. It also attacks the young shoots.

12 Bee-parasite Beetle (*Stylops melittae*). An insect now classified, not as a beetle, but in an order of its own, the Strepsiptera. The picture shows a female with its body protruding from the abdomen of its host, an *Andrena* bee (p. 158). The tiny, short-lived male flies with a humming noise early on hot summer mornings, with a side-to-side motion, about places frequented by the bees, mainly in southern England. The white, blind, larva-like female never leaves the host. She produces not eggs but tiny, active larvae which escape into the bees' nest. There, they bore into the bee grubs and change to a maggot-like form. They do not seem to harm the bees much, one bee being able to support up to three parasites. The large holes the escaping males leave in the bee's tail soon close up.

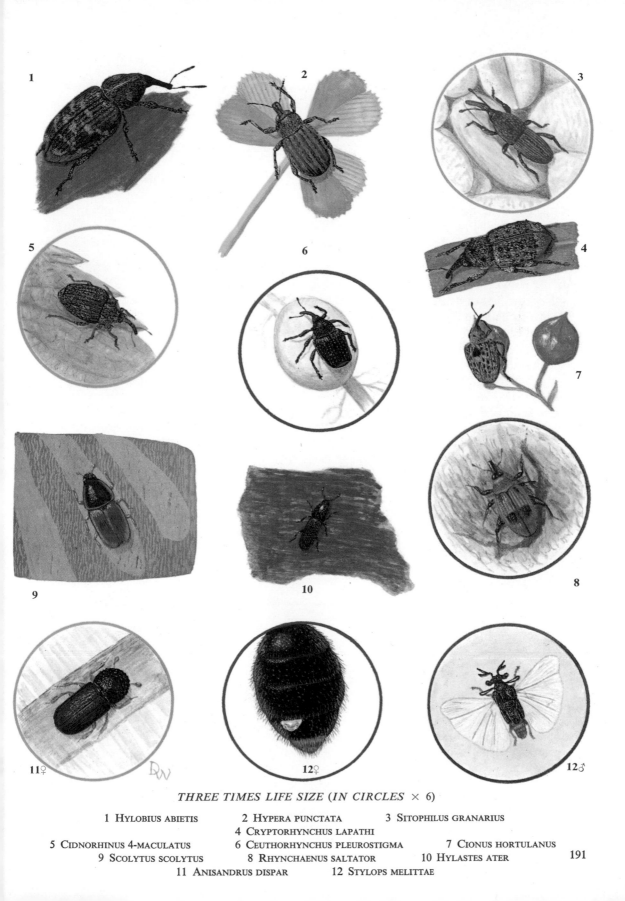

THREE TIMES LIFE SIZE (IN CIRCLES × 6)

1 HYLOBIUS ABIETIS 2 HYPERA PUNCTATA 3 SITOPHILUS GRANARIUS
4 CRYPTORHYNCHUS LAPATHI
5 CIDNORHINUS 4-MACULATUS 6 CEUTHORHYNCHUS PLEUROSTIGMA 7 CIONUS HORTULANUS
9 SCOLYTUS SCOLYTUS 8 RHYNCHAENUS SALTATOR 10 HYLASTES ATER
11 ANISANDRUS DISPAR 12 STYLOPS MELITTAE

CLASSIFICATION OF BRITISH INSECTS

Insects belong to the group of invertebrate animals called Arthropoda, which are divided into four classes — Insecta (the insects); Arachnida (spiders, mites, ticks, and scorpions); Myriapoda (centipedes and millipedes); and Crustacea (crabs, lobsters, woodlice, and others).

Of all the animal species in the world it is estimated that about five out of six are insects. Nearly a million species have been named, and more are being discovered every year. It is probable that there are in existence about three million species or even more. In Britain there are about 20,000 named species, with more being added each year. It is not surprising that among so many species, relatively few have been given English names, and of these even fewer are well enough known for their names to be in general use. So most are known only by their scientific names, which are derived from Latin or Greek, and are recognized by all entomologists all over the world.

The scientific name, as with all plants and animals, consists of two parts — a generic and a specific name. For example, the Silver-washed Fritillary butterfly (*see* p. 44) is called *Argynnis paphia*: the first or generic name is shared with other closely related butterflies such as the Dark Green Fritillary and the High Brown Fritillary, which belong to the same genus *Argynnis*; the second is the specific name, and, though other animals may also have the same specific name, none have both the same generic and specific name. Some species have a third subspecies name, which indicates a distinct local race. For example, the race of the Oak Egger moth (p. 91) occurring in northern Britain, *Lasiocampa quercus callunae*, is distinguished from the typical species by the word *callunae*. Also where there are well-defined varieties or forms of a species, these may be shown by an additional name — for instance, *Polygonia c-album* var. *hutchinsoni*, which refers to a variety of the Comma Butterfly (p. 48), first identified by Hutchinson. In specialist books scientific names are followed by Linn. or Thunb. or some other abbreviated name, not in italics. This refers to the scientist who first described and named the species.

Related genera are grouped together into families. Where a group of genera within a family have certain common characteristics, they may be described as a subfamily. Families are grouped into orders (in which there may also be suborders). According to the latest classifications there are twenty-five different orders of British insects, though some authorities recognize only twenty-two, grouping some orders together. This book has followed the classification used by Prof. V. B. Wigglesworth in his book *The Life of Insects* (1964), though with occasional modifications for the sake of convenience.

STRUCTURE OF INSECTS

Insects, which form one class of the invertebrate animals, Arthropoda, have bodies made up of segments, one behind the other, some of which carry appendages such as legs or wings. An insect's body has an outer skeleton or horny covering to which the body is attached by muscles. This skeleton does not grow, so the insect can grow only by shedding or 'moulting' the skeleton at intervals (*see* p. 199). It is jointed, like a suit of armour, to allow the insect to move, and it is made chiefly of a substance called 'chitin' which is exceedingly hard and durable.

An insect's body consists of a head, thorax, and abdomen (Fig. 1). The head carries a pair of sense organs, called antennae, which may vary from the tiny appendages of a Hover-fly (p. 131) to the enormously long antennae of a Longicorn Beetle (p. 185), four or five times as long as its body. The mouthparts vary according to the way the insect feeds — whether it bites, chews, or sucks. There are three parts — a pair of mandibles (upper jaws), a pair of maxillae (lower jaws), and the labrum and labium (upper and lower lip). In the bugs the mouthparts form a rostrum, or beak, for piercing (*see* p. 195), and in other insects, including certain flies and moths, they form a proboscis or sucking tube. Insects normally have two large compound eyes, which are very complicated and efficiently designed for seeing quick movements. As well, they have simple eyes or ocelli, usually in a group of three, which consist of a single lens and are probably sensitive to changes in light.

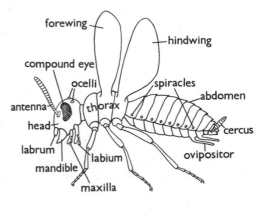

Fig. 1

The thorax is made up of three segments, each normally carrying a pair of legs, and the second and third carry a pair of wings. In some insects, the two-winged flies, for example, the hindwings are reduced to stubs, called halteres (*see* p. 196). The wings are strengthened by a network of veins, each type of insect having a characteristic pattern. The wings of some insects — butterflies and moths, for example — are covered with scales or hairs, and the forewings of others, such as beetles, serve as a protective covering for the hindwings, which are used for flight. The legs are made up of five main segments (Fig. 2), which vary from one insect to another according to the use made of them — for running, jumping, digging, or swimming.

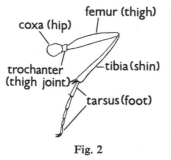

Fig. 2

The abdomen is usually divided into ten segments, the last three or four carrying the sex organs. The male may carry a pair of claspers for grasping the female during mating, and the female has an ovipositor or egg-laying apparatus, which with most insects is hidden from view, but with some, such as Wood-wasps (p. 143), is very long. With some insects the ovipositor also serves as a sting or as a piercing tool.

Insects breathe very differently from most animals. Instead of lungs, they have a network of breathing tubes (trachae) which carry oxygen to all parts of the body. The oxygen is absorbed from the air or water through a series of air-holes (spiracles) along the body. An insect's 'heart' is a long tube equipped with valves for pumping the 'blood' round the body. The 'blood', which fills the whole body cavity, and carries nutritious substances to the tissues, carries virtually no oxygen and no haemoglobin, and is usually greenish or yellowish, instead of red. The nervous system consists of a long, double, nerve-cord running the whole length of the body along the underside. In each segment there are nerve centres called ganglia, which are like local 'brains', controlling the activities of that part of the body. To some extent they will continue to function even if cut off from the main brain in the head. For instance, the hind-part of a wasp will continue to sting for a time after the head has been cut off, or the head to eat after the abdomen has been removed. An insect can store food in groups of cells in its abdomen, and these can be used up when the insect cannot eat — for example, during hibernation or during the pupal stage.

ORDERS OF BRITISH INSECTS

The twenty-five orders of British Insecta are grouped into two sub-classes, the Apterygota (1 – 4) and the Pterygota (5 – 25). The latter is again sub-divided into the Exopterygota (5 – 15) and the Endopterygota (16 – 25).

The Apterygota, or wingless insects, are small, primitive insects which have no complete metamorphosis, the young stages differing little from the adult stage, except in size. The Pterygota are typically winged insects, though there are some wingless forms. Division I of these, the Exopterygota, are insects in which the wings develop externally in the larvae, which are usually called 'nymphs'. These resemble the adults in a general way, some more and some less, and undergo only an incomplete metamorphosis, for there is no pupal stage. Division II, the Endopterygota, are insects in which the wings develop within the body of the larva. The larvae, in general, look quite unlike the adults, and they undergo a complete metamorphosis. During the pupal or resting stage the transformation from larva to adult takes place (*see* p. 198).

The arrangement of this book is according to the generally accepted order, with some variations for convenience.

SUB-CLASS I. Apterygota (p. viii)

Order 1   Thysanura (Three-pronged Bristle-tails), 9 British species (p. viii). Details of this and the following three orders are given on p. viii.

Order 2   Diplura (Two-pronged Bristle-tails), 12 British species.

Order 3   Protura (Proturans). 12 British species. These are the only insects without antennae.

Order 4   Collembola (Springtails). 304 British species. These insects differ from others in having abdomens with six instead of the usual ten or eleven segments.

SUB-CLASS II. Pterygota — Division I,
Exopterygota (pp. 2 – 35)

Order 5   Odonata (Damselflies and Dragonflies), 45 British species (pp. 2 – 7). These insects have two pairs of net-veined wings of similar size and shape, with a dark mark or 'stigma' near the tip of each. They have long, slender abdomens, large heads, very large compound eyes, and little, hair-like antennae. They catch insects in flight and hold them to their mouths with special attachments on their spiny legs. The nymphs are aquatic. The more slender Damselflies (Zygoptera) have a weak flight and rest with their wings folded over their abdomens. Dragonflies (Anisoptera) rest with their wings rigidly outspread.

These insects have a unique way of mating. The male, in preparation, curves his body to transfer his sperm from the tip to a special opening under the second and third segments of the abdomen. Then he grasps the female by the back of her head or neck with his tail claspers. The female curves up her abdomen so that the tip enters the opening in the male's body and receives the sperm (Fig. 3). A pair can often be seen flying 'in tandem' in this position.

Order 6   Ephemeroptera (Mayflies), 46 British species (p. 8). The name comes from a Greek word meaning 'living a day', for the adults only live for one to four days. The nymphs are aquatic, with gills. The adults have very short antennae, two or three long tail-filaments, and transparent wings, of which the forewings are larger than the hind. When resting, they hold their wings over their backs, as butterflies do.

Fig. 3
Mating Damselflies

194

Order 7 Dictyoptera (Cockroaches and Mantids), 9 British species (p. 10). This order is closely related with Order 10, with which it is often included. These insects, however, have their feet (tarsi) divided into five segments instead of only three or four. They have biting and chewing mouthparts (Fig. 4a). The Cockroaches (Blattodea) are well-known British insects, but the Mantids (Mantodea) have no British representative, though they are common on the Continent.

Order 8 Plecoptera (Stoneflies), 34 British species (p. 8). These insects in some ways resemble those in Orders 16 and 17, but they have hindwings broader than the forewings and long 'tails' (cerci) at the end of the abdomen. Also they rest with their wings folded flat instead of held tent-like over their backs. The nymphs are aquatic, usually with gills.

Order 9 Dermaptera (Earwigs), 7 British species (p. 8). These are described on page 8.

Order 10 Orthoptera (Crickets, Grasshoppers, etc.), 30 British species (pp. 10 – 17). This order is now often called Saltatoria. It is sub-divided into the Ensifera (Bush-crickets, Crickets, Mole-crickets — pp. 10 – 13) and the Caelifera (Grasshoppers, Locusts, and Groundhoppers — pp. 14 – 17). The former have long, thread-like antennae and produce their 'song' by rubbing specially modified parts of their forewings together. The latter have short antennae and usually 'sing' or stridulate by rubbing tiny pegs on the inside of the femur of the hind-leg against thickened veins on the forewing.

Order 11 Psocoptera (Book-lice), *c.* 87 British species (p. 18). Most of these have wings and live out-of-doors, but the best-known are wingless and live indoors.

Order 12 Mallophaga (Biting-lice), 514 British species (p. 18). Small, flattened, wingless insects, with short antennae, mostly living as parasites on birds, and described on p. 18.

Order 13 Anoplura (Sucking-lice), 25 British species (p. 18). These lice used to be classed as the Siphunculata, and they and the Mallophaga considered as sub-orders of the Anoplura. Now they are each considered separate orders, the one with sucking and the other with biting mouthparts.

Order 14 Thysanoptera (Thrips), *c.* 160 British species (p. 18). These tiny insects also have mouthparts adapted for piercing and sucking, but most attack plants, although some attack other insects. They have wings and can fly.

Order 15 Hemiptera (Bugs), *c.* 1630 British species (pp. 20 – 35). A very large order in which the mouthparts are specially modified into a 'beak' (rostrum). This contains four needle-like rods, used for piercing either plants or animals and sucking juices or blood (Fig. 4). Bugs vary a great deal and are divided into two sub-orders. The Heteroptera (pp. 20 – 31), which have membranous tips to their otherwise leathery forewings, rest with their wings held flat over the body, the fore-wings protecting the completely membranous hindwings. The Homoptera (pp. 32 – 35), in those cases where they have wings at all, rest with these held in a tent-like manner over the back.

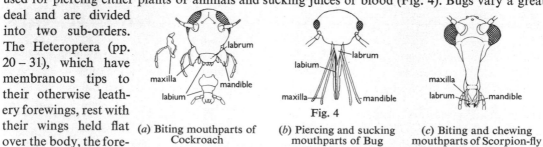

Fig. 4

(*a*) Biting mouthparts of Cockroach   (*b*) Piercing and sucking mouthparts of Bug   (*c*) Biting and chewing mouthparts of Scorpion-fly

SUB-CLASS II Pterygota — Division II,
Endopterygota (pp. 36 – 191).

Order 16   Megaloptera (Alder-flies and Snake-flies), 6 British species (p. 36). These are very much like the Lacewings, Order 17, and are still often grouped with them. These and all the remaining orders have a full metamorphosis; egg, larva, pupa, and adult (*see* p. 198).

Order 17   Neuroptera (Lacewings), 54 British species (p. 36). These used to be grouped with the Megaloptera, and the small differences between them are described on p. 36.

Order 18   Mecoptera (Scorpion-flies and Moss-flies), 4 British species (p. 36). These have beak-like faces with biting and chewing mouthparts (Fig. 4c).

Order 19   Trichoptera (Caddis-flies), 193 British species (p. 38). The name means 'hairy-winged', and these insects differ from moths in having hairs instead of scales on the wings. The mouthparts of the adults are atrophied, and most species cannot feed.

Order 20   Lepidoptera (Butterflies and Moths), *c.* 2,190 British species (pp. 40 – 121). This order can be divided into the Butterflies (Rhopalocera), which have clubbed antennae, all fly by day, and usually rest with their wings closed over the back, the forewings almost completely covered by the hindwings; and the Moths (Heterocera), which often have thicker bodies, and antennae of various shapes, but not clubbed. They rest with their wings in varied positions, according to species, and, though most fly by night, many are day fliers.

Butterflies and Moths are also divided into two sub-orders (or by some authorities, three) on more scientific principles. The Homoneura, by far the smaller group, contains rather primitive insects such as the Swifts (p. 114), with the same vein pattern on both fore- and hind-wings. Instead of the more usual device for coupling the fore- and hind-wings together in flight, they have either no coupling apparatus or a small flap projecting from the hind edge of the forewing which looks as though it should, but appears to fail to, make a coupling with the front edge of the hindwing (Fig. 5a). The Heteroneura have a different vein pattern on the fore-and-hind-wings, a well-developed

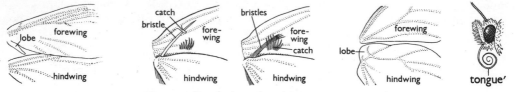

Fig. 5.   **Wing-coupling devices of Lepidoptera**
    (*a*) Homoneura   (*b*) Frenulum of male and female moth   (*c*) Butterfly   Fig. 6

proboscis for sucking nectar, which, when not in use, is coiled (Fig. 6), and a coupling device for the wings called a frenulum. This consists of one or more bristles on the leading edge of the hindwing, which is gripped by a catch on the underside of the forewing. The device differs slightly between the two sexes (Fig. 5b). Butterflies have no frenulum, but instead a lobe on the leading edge of the hindwing which locks beneath the overlapping forewing (Fig. 5c).

The larvae of Lepidoptera, usually called caterpillars, can secrete silk from a gland in the head, with which they spin cocoons to protect their pupae or chrysalises, attach the pupa to an object, or make larval shelters. The adults of most species can suck food, but cannot bite or chew.

Order 21   Diptera (Two-winged Flies), *c.* 5,200 British species (pp. 122 – 141). The main characteristic of these insects is that they have only one pair of wings. (A few species are wingless.) The hindwings are represented only by small balancer organs, called 'halteres', which are important in regulating stable flight. Flies have sucking mouthparts, and some can pierce the skin of animals in order to suck their blood. The larvae are legless.

There are three sub-orders: the Nematocera are rather primitive, fragile insects, usually with long antennae, such as Crane-flies and Mosquitoes (pp. 122 – 125); the Brachycera are more powerful flies such as Soldier Flies, Horse Flies, and Bee Flies, which have short antennae (pp. 126 – 129); the Cyclorrhapha also have very short antennae, the larvae have no separate heads, and the white pupae are protected by hard, dark, egg-shaped puparia, formed from the final larval skin. They include a wide variety of flies, many of them pests to human beings, such as Blow Flies and Warble Flies (pp. 130 – 141).

Order 22 Siphonaptera (Fleas), 47 British species (p. 18). These have much in common with the two-winged flies, but are described in this book with other parasitic insects, such as Lice.

Order 23 Hymenoptera (Saw-flies, Ichneumon-flies, Ants, Bees, and Wasps), *c.* 6,190 British species (pp. 142 – 163). This is the largest order of British insects. Most species have two pairs of transparent, membranous wings, the hind pair, which are the smaller, being linked to the forewings in flight by rows of tiny hooks. The females have ovipositors at the tip of the abdomen which, in some species serve as drills, saws, or stings (Fig. 7). The males cannot sting. They usually have biting mouthparts, though those of some, for example, bees, are adapted for sucking.

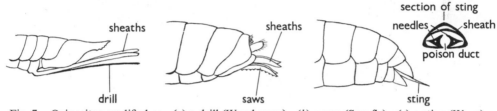

Fig. 7. Ovipositor modified as (*a*) a drill (Wood-wasp) (*b*) a saw (Saw-fly) (*c*) a sting (Wasp)

There are two sub-orders. The Symphyta, the Wood-wasps and Saw-flies (pp. 142 – 145), have no 'wasp-waist' between thorax and abdomen. The females have very long, and in some species saw-like ovipositors (Figs. 7a, b). The larvae are plant-feeding and caterpillar-like. They have legs— three pairs of true legs on the thorax and at least six pairs of false legs on the abdomen (more than those of Lepidoptera). The Apocrita (pp. 146 – 163) all have 'wasp-waists', and the larvae are grub-like and legless. They are sub-divided into the Parasitica (pp. 146 – 149), the larvae of which are mostly parasites on other insects or plants, and the Aculeata (pp. 150 – 163), in which the female ovipositor is modified as a sting (Fig. 7c) and in which the larvae are helpless and depend on care by the adults. Many of these have evolved a highly-developed social organization of special-ized duties.

Order 24 Coleoptera (Beetles and Weevils) *c.* 3,690 British species (pp. 164 – 191). The members of this very large order of four-winged insects have their forewings modified to form shell-like wing-cases, called 'elytra'. These not only protect the soft abdomen but also the membranous hindwings which are folded beneath them when the insect is at rest. When the beetle flies, the elytra are raised, and the hindwings only are used in flight. In some species the wings have become either degenerate or absent.

The order is divided into the Adephaga or carnivorous Beetles (pp. 164 – 169) and the Polyphaga or omnivorous Beetles (pp. 170 – 191). The Weevils, which have a beak-like extension to the head, the 'rostrum', belong to one large family, in which there are more species than in any other family of animals in the world.

Order 25 Strepsiptera (Stylops), 17 British species (p. 190). This order is sometimes classified with the Coleoptera. The wingless females and larvae are parasitic on Bees and Wasps. The males have hindwings only, the forewings having become modified, like those of the Diptera.

METAMORPHOSIS OF INSECTS

Insects, like most other members of the large group of invertebrate animals, the Arthropoda, pass through changes of structure during their lives which are called their metamorphosis. These animals, in contrast to vertebrate animals, have their skeletons on the outside, and as they grow, they cast off outgrown skeletons and larger ones underneath harden in their place. Each casting is called a 'moult'. Apart from some exceptions, these animals all start as eggs which hatch into larvae. With a few insects, such as Silverfish and Firebrats (p. 1), the change in outward form between larva and adult is so little that it can hardly be called a metamorphosis. The Exopterygota division of insects (p. 194) have what is called an incomplete metamorphosis, the larvae not differing from the adults to a great extent, and there being no resting or pupal stage. The Endopterygota division (p. 195) have a complete metamorphosis consisting of four distinct stages—egg, larva, pupa, and adult. It is simplest to describe this first.

Typical insects which undergo a complete metamorphosis are butterflies and moths, two-winged flies, beetles, and bees, wasps, and ants. Fig. 8 shows the four stages of metamorphosis of the

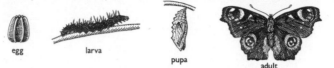

egg   larva   pupa   adult

Fig. 8.   Metamorphosis of Peacock Butterfly

Peacock Butterfly. The larvae vary a great deal from one Order to another, according to their mode of life. Those that have to search for their food, such as those of most butterflies and moths, and ground beetles, possess legs — usually three pairs on the first three segments of the body (which later become the six legs on the adult's thorax), and further pro-legs on the hind segments, which disappear in the adult. Those which live surrounded by food, such as the larvae of bees, flies, and some beetles, are legless.

The larva is concerned with eating and growing. The number of moults vary with the species, from two or three up to as many as fifty. When it is fully grown, the larva seeks a suitable place to pupate. It may spin a cocoon of silk, mixed or not with other materials; it may form a pupal cell of soil, plant tissue, or other substance; it may retain the larval skin of the final moult, which hardens and darkens into a protective puparium around the pupa; it may merely attach itself to some object, without any outside protection.

The pupal stage is one of rest, during which the insect is inactive and does not feed. During this period, its body is more or less completely broken down and reconstructed in the adult form. The position and shape of the antennae, eyes, mouthparts, legs, and wings are formed. When the metamorphosis is complete and the adult ready to emerge, it breaks out of the pupa, climbs to some suitable vantage point to rest, excretes a considerable amount of fluid — the waste materials from the changes which have taken place in the pupa, expands and dries its wings, and finally takes flight.

Typical insects which undergo an incomplete metamorphosis are grasshoppers, dragonflies, and bugs. With these insects the change from larva (usually called nymph) to adult is more gradual and includes no pupal stage. The nymphs are never very dissimilar from the adult, and they gradually grow more alike, the wings developing externally and lengthening with each moult, until after the last moult, the insect is mature (*see* Fig. 9).

Fig. 9.   Stages in incomplete metamorphosis of Grasshopper

PROTECTION FROM ENEMIES

Insects are the natural food of other insects, birds, and many other animals, and vast numbers are eaten. Their powers of reproduction are also vast, but even then they might not survive had they not evolved through the centuries various methods of avoiding attack. These methods, however ingenious and intelligent they seem, are never purposeful action on the part of the insect. A type of behaviour which happens to deceive an enemy becomes in time an instinctive behaviour pattern, and insects with a successful coloration survive and breed, while those without are destroyed.

Camouflage, or cryptic (secret) coloration, is one of the most effective methods of defence, especially for insects which are sought after as food. The most common kind of camouflage is a close resemblance to the natural surroundings — the foodplant or the tree, wall, or grass on which the insect usually rests during the day. There are endless examples of this among butterflies and moths. The Merveille du Jour (p. 79) looks like a patch of lichen when resting by day on a lichen-covered oak-tree; the Thorn Moths (p. 107) closely resemble shrivelled leaves, and their cater-pillars resemble in colour, shape, and posture the twigs of a tree; the Chinese Character Moth (p. 69), when resting on a leaf, looks like a bird-dropping, and many beetles and weevils are easily mistaken for bits of earth, stones, or seeds, so long as they stay still. The Grayling Butterfly (p. 43), whose underwings have excellent cryptic coloration, adds to its concealment by leaning over to one side so as to reduce the length of its shadow. Successful camouflage is also achieved by a pattern of colours which among leaves or grass breaks up the outline of the insect and makes it difficult to detect. The disruptive colours of the Common Field Grasshopper (p. 15) make it quite invisible among dead and living grasses, and even large, brightly-coloured larvae such as the Privet Hawk-moth (p. 61) become invisible among the leaves.

With the growth of large industrial areas in the last 100 years, a new form of camouflage has been evolving, which we call industrial melanism: black or very dark varieties of insects frequenting these areas have been appearing. For instance, in 1850 a black variety of the Peppered Moth (p. 111) appeared in Manchester. At first it was very rare, but gradually it increased in numbers until by about 1900 99% of the Peppered Moths in this area were black. They spread to other industrial areas, and now are by far the more common, the light ones only being common in areas such as south-west England. Obviously in smoky districts where tree-trunks, leaves, walls, and so on become sooted, a light insect is conspicuous and is soon captured, whereas a darker one escapes notice and lives to breed and increase. There are many examples among Lepidoptera of this development, and also of some other insects such as the Cuckoo Bumble-bee (p. 163). The abundance of blackish Common Field Grasshoppers in the suburbs of London may be due to the same change to suit a polluted environment.

Another use of colour for protection is called 'flash' coloration. The Red and Yellow Underwing Moths (pp. 87, 73), for example, have drab, camouflaged forewings and bright (flash) colours on the hindwings. A predator pursues this bright-coloured insect, which seems suddenly to disappear. In fact, the moth has alighted and immediately folded its bright hindwings beneath the forewings, and so is no longer a bright insect at all. The conspicuous 'eyes' on the wings of many butterflies and moths also serve as a distraction. If the predator strikes before the insect pursued has had time to 'disappear', the predator will strike at the 'eyes' instead of the far more vulnerable body, and the insect will survive.

Many insects use their 'eyes' and other striking features aggressively to frighten off an enemy. The Peacock Butterfly (p. 49), for example, looks like a dead leaf when at rest, but if attacked, it rapidly opens and shuts its wings, displaying its four great eyespots and probably dismaying its enemy for long enough to allow it to escape. Many larvae employ this kind of aggressive use of their colours or other features. The Elephant Hawk-moth and the Puss Moth larvae (p. 65), if disturbed, assume a terrifying stance likely to deter a predator. The male Stag Beetle (p. 183) rears on its hind-legs and opens its huge antler-like mandibles, which, in fact, are not as formidable as they look.

Some insects use their colours to draw attention to themselves. Such insects are usually poisonous or at least unpalatable, and their colour pattern warns predators that the last time they attacked an insect of this pattern they disliked it. Cinnabar (p. 95) and Burnet (p. 113) Moths carry these warning colours, as do Ladybirds (p. 177) and Cardinal Beetles (p. 179), as well as Wasps, which warn enemies that they can sting. The gaudy Tiger Moths (p. 95) display their bright colours to the fullest extent and also produce a warning scent, which is a complement to warning colours with many insects. Many warningly-coloured larvae also possess poisonous or irritating hairs. Other insects, such as ants (p. 150), can squirt poison in the face of an enemy.

Many insects, which are not themselves poisonous, achieve protection by mimicking, both by colour and behaviour, those insects which are. A bird would hesitate to attack a Hover-fly (p. 131) in case it should turn out to be a bee or a wasp, and Wasp Beetles (p. 185), Bee Hawk-moths (p. 63), Hornet Moths (p. 115), and Bee Chafers (p. 183) gain protection for the same reasons. A black-and-yellow banded pattern, shared by many wasps and also by many other insects, becomes well-recognized as warning colours, and so are very effective.

In spite of these ingenious methods of defence, some predators specialize on these 'protected' insects. Bee-eaters, Shrikes, and Honey Buzzards eat bees and wasps, and Cuckoos eat many warningly-coloured caterpillars. On the whole, however, these methods of protection are effective or they would not have continued, and, as is shown by the growth of industrial melanism, protective devices continue to evolve to meet new conditions.

SOURCES OF FURTHER INFORMATION

BOOKS   A: suitable for beginners; B: for identification and general reading; C: books of reference.

General
A   IMMS, A. D., *Insect Natural History*. 2nd edn., 1956. Collins. 35/–
A   NEWMAN, L. H., *Instructions to Young Naturalists: Insects*. 1956. Museum Press. 12/6
A   OLDROYD, H., *Insects and their World*. 2nd edn. 1966. British Museum (Natural History). 7/6
A   BROWN, E. S., *Life in Fresh Water*. 1955. O.U.P. 12/6
B   ROYAL ENTOMOLOGICAL SOCIETY, *Handbooks for the Identification of British Insects*.
     Details obtainable from the Registrar, R.E.S. 41 Queen's Gate, London S.W.7.
B   WIGGLESWORTH, V. B., *The Life of Insects*. 1964. Weidenfeld and Nicolson. 55/–
C   IMMS, A. D., *A General Textbook of Entomology*. 9th edn. revised by O. W. Richards and R. G. Davies.
     1957. Methuen. 84/–
C   OLDROYD, H., *Collecting, Preserving, and Studying Insects*. 1963. Hutchinson. 30/–

Coleoptera
B   LINSSEN, E. F., *Beetles of the British Isles* (2 vols.). 1959. Warne. 31/6 each

Diptera
B   COLYER, C. N. & HAMMOND, C. O., *Flies of the British Isles*. Revised edn. 1951. Warne. 35/–

Hemiptera
B   SOUTHWOOD, T. R. E. & LESTON, D., *The Land and Water Bugs of the British Isles*. 1959. Warne. 37/–

Hymenoptera
A   COOPER, B. A., (Ed.) *The Hymenopterist's Handbook*. 1943. Amateur Entomologist's Society. 10/6
B   STEP, E., *The Bees, Wasps, Ants, and Allied Insects of the British Isles*. New edn. in preparation. Warne. 15/–

Lepidoptera
B   FORD, E. B., *Butterflies*. 3rd edn. 1957. Collins. 35/–
B   FORD, E. B., *Moths*. 1955. Collins. 35/–
B   NEWMAN, L. H., *The Complete British Butterflies in Colour*. 1968. Edbury Press and Michael Joseph. 35/–
B   SOUTH, R., *The Butterflies of the British Isles*. New edn. in preparation. Warne. 21/–
B   SOUTH, R., *The Moths of the British Isles* (2 vols.). 1961. Warne. 37/– each

Odonata
B   CORBETT, P. S., LONGFIELD, C. & MOORE, N. W., *Dragonflies*. 1960. Collins. 42/–

Orthoptera
B   RAGGE, D. R., *Grasshoppers, Crickets and Cockroaches of the British Isles*. 1965. Warne. 44/–

JOURNALS
Entomologist's Gazette, pub: by E. W. Classey Ltd., 353 Hanworth Road, Hampton, Middlesex.
The Entomologist's Monthly Magazine, pub: by Dr. B. M. Hobby, 7 Thorncliffe Road, Oxford.
The Entomologist's Record. L. Parmenter, 'Woodside', Pinewood Road, Ferndown, Dorset.
The Entomologist, pub: by British Trust for Entomology.

SOCIETIES
The Royal Entomological Society, 41 Queen's Gate, London S.W.7.
The British Trust for Entomology, 41 Queen's Gate, London S.W.7.
The Amateur Entomologists' Society, Sec. R. D. Hilliard, 18 Golf Close, Stanmore, Middlesex.
These all publish their own journals and other publications.